THE
INTERNATIONAL SERIES
OF
MONOGRAPHS ON PHYSICS

GENERAL EDITORS

W. MARSHALL   D. H. WILKINSON

MAGNETIC RESONANCE IN METALS

BY

J. WINTER

OXFORD
AT THE CLARENDON PRESS
1971

Oxford University Press, Ely House, London W. 1

GLASGOW NEW YORK TORONTO MELBOURNE WELLINGTON
CAPE TOWN IBADAN NAIROBI DAR ES SALAAM LUSAKA ADDIS ABABA
DELHI BOMBAY CALCUTTA MADRAS KARACHI LAHORE DACCA
KUALA LUMPUR SINGAPORE HONG KONG TOKYO

PRINTED IN GREAT BRITAIN
AT THE UNIVERSITY PRESS, OXFORD
BY VIVIAN RIDLER
PRINTER TO THE UNIVERSITY

PREFACE

THE first experiments on nuclear and electronic resonance were performed in the mid-1940s. The aim was to gain a better understanding of the magnetic resonance properties themselves. As will be discussed in this book, the coupling between the nuclear moments and the other degrees of freedom of the system usually called 'the lattice' is very weak. In a nuclear magnetic resonance experiment we therefore study the motion of a nearly-free magnetic moment. (This statement is only an approximation because there are interactions between the nuclear moments.) The properties met in an electronic spin resonance experiment are quite different. The occurrence of electronic magnetism is a less general property of matter; it occurs only in substances containing unpaired electrons. The first experiments were done on ionic crystals containing elements of the iron group; the aim of these studies was to investigate the structure of the ground state of these ions and the work was merely an extension of classical atomic spectroscopy.

In considering nuclear resonance properties, it can be said that, because the coupling between the nuclear spin system and the lattice is weak, the nuclear moments provide a very good probe for studying local magnetic fields in condensed matter (the nuclear moments are also coupled to the electric field gradients). The motion of the nuclear moments will not perturb the properties of the lattice. A large number of experiments have been performed using nuclear moments to test the magnetic properties found in condensed matter. One of the best examples is provided by the studies of nuclear magnetic resonance in metals.

The study of electronic spin resonance in metals started at Berkeley in the mid-1950s. Rather surprisingly, after some very interesting initial results and an important theoretical contribution, due to Dyson, no important results were reported for about ten years. This is presumably explained by the existence of serious technical difficulties. More recently the discovery of a new technique has resulted in a large number of new results in this area. We can mention here that the properties of the electron spin resonance of conduction electrons are very different from the properties of the localized spins met in ionic crystals and are also very different from the nuclear resonance properties. This is a consequence of the non-localization of the spin carriers and of

Fermi–Dirac statistics. In some cases one finds electronic spins in a situation intermediate between purely localized spin and conduction-electron spin. This situation is met in semiconductors where the electrons have wave functions extending over very large orbits. In such substances the properties are strongly dependent on various parameters such as the temperature, the concentration of impurities, and so on. These experiments are interesting but will not be discussed here.

In this book we have tried to treat the electronic and nuclear properties on the same footing. The nuclear magnetic resonance properties are strongly affected by the magnetic fields produced by the magnetization of the conduction electrons. In some cases (such as the Overhauser effect) a perturbation of one of the systems produces drastic changes in the properties of the other system. In fact if quadrupolar and orbital hyperfine couplings are neglected the nuclear moments are coupled with the lattice by their coupling with the spin magnetization of the conduction electrons.

It is often important to add to the nuclear or electronic resonance experiments the result of experiments using other techniques. For example, it is well known that the discovery of the Mössbauer effect has been useful in studies of nuclear resonance in ferromagnetic metals. In other situations a measurement of the static susceptibility or of the specific heat will be helpful in addition to nuclear resonance measurements in gaining a better understanding of the properties of metals.

In this book nuclear magnetic resonance is studied first. An introductory chapter deals with principles of nuclear magnetism, and in the following chapter the hyperfine interactions, which couple the nuclear spin system with the lattice, are discussed. These two chapters are brief and are intended only to remind the reader about these subjects. The third chapter is a review of some properties of metals, more especially the magnetic properties. The following chapters contain a detailed discussion of the theory of, and the experimental work on, nuclear resonance in metals. The discussion includes the quadrupolar interactions in metals, nuclear resonance in alloys, the properties met in transition metals or alloys, and nuclear resonance in superconductors. The last chapter is devoted to a study of electronic resonance. The later chapters deal with subjects that are far from being completely understood. There is a constant and large flow of publications in these fields and the aim in these chapters is not to review all the known results but to give to the reader the knowledge necessary to understand the literature.

I am greatly indebted to Professor A. Abragam, who not only suggested the writing of this monograph, but who initiated me into nuclear
and electronic resonance properties. I am very pleased to thank all my
friends in the magnetic resonance laboratory at Saclay for many discussions on the problems discussed in this book. I cannot give all their
names but would like to mention particularly Dr. A. Landesman and
Dr. M. Goldman. I would especially like to thank Dr. M. Odhenal and
Dr. F. I. B. Williams, who read the manuscript in detail and suggested
many improvements. Finally I am greatly indebted to Mme M. Porneuf
and Mme F. Lefèvre for their great help with the preparation of the
manuscript.

J. W.

June 1970

CONTENTS

NOTATION TABLE

$C_{\mathbf{k},\sigma}^{+}, C_{\mathbf{k},\sigma}$	creation and destruction operators
eq	electric field gradient
$g(E)$	density of electronic states
$\mathbf{H}_0$	external static field
$\mathbf{H}_e$	hyperfine field
$\hbar\omega_0$	electronic Zeeman energy
$\hbar\omega_n$	nuclear Zeeman energy
$\mathbf{I}$	nuclear spin
I_n	intensity of the nuclear resonance signal
K	Knight shift
$\mathbf{l}_e$	electronic orbital moment operator
M_s or $\mathbf{M}$	electronic spin magnetization
$\mathbf{m}$	nuclear moment
N_+, N_-	populations of the nuclear sub-levels
$n(E)$	occupation number
Q	electric quadrupole moment
$\mathbf{S}(R)$	local spin density operator
s	electronic saturation factor
$\mathbf{s}_e$	electronic spin operator
T_1	longitudinal nuclear relaxation time
β	Bohr magneton
γ_n	nuclear gyromagnetic ratio
χ_n^0	static nuclear susceptibility
χ_n', χ_n''	time variable nuclear susceptibilities
χ_s	electronic spin static susceptibility
χ_s^F	static spin susceptibility for a free electron gas
χ_s', χ_s''	time variable electronic spin susceptibilities
$\chi_s(q), \chi_s(\omega, q)$	generalized spin susceptibilities
ω_c	cyclotron resonance frequency

SHORT REVIEW OF THE PROPERTIES OF NUCLEAR MAGNETIC RESONANCE

I N this chapter, we shall review very briefly the basic properties of nuclear magnetic resonance. Since full details of this subject are contained in References 1–3, only some of the more important features will be given here, without calculations or justifications.

1. Some considerations of the order of magnitude of nuclear Zeeman energy

A large number of nuclei have in their ground state a non-zero angular momentum $\hbar\mathbf{I}$ and consequently a magnetic moment. The magnitude of this moment is given by the equation

$$\mathbf{m} = \gamma_n \hbar \mathbf{I}, \tag{1.1}$$

where γ_n is the nuclear gyromagnetic ratio. This quantity is typically 1000 times smaller than an electronic gyromagnetic ratio.

In a static external field $\mathbf{H}_0$, the nuclear Zeeman energy is given by the equation

$$E = -\mathbf{m}.\mathbf{H}_0. \tag{1.2}$$

It is worth while to discuss the order of magnitude of this energy. In a magnetic field of 10^4 G, when converted into frequency units, it is of the order of 10 MHz (in the case of hydrogen nuclei, the value is 42·58 MHz). It is therefore extremely small compared to all the other possible energies found in condensed matter, and this fact has several important consequences that will be reviewed now.

1.1. *Static nuclear susceptibility*

As nuclear moments are three orders of magnitude smaller than electronic moments, the interactions between nuclear moments are expected to be 10^6 times smaller than the interactions between electronic moments. Moreover, the nuclei are well localized and there is no exchange interaction due to the overlap of the nuclear wave functions (apart from the important exception of ^{3}He where a nuclear exchange interaction exists even in the solid phases). As the interactions are purely dipolar their magnitude is found to be of about 20 Hz to 2 kHz

B

in frequency units (or 10^{-6} to 10^{-7} K in temperature units). Throughout the usual temperature range, the system of nuclear moments behaves as an ideal paramagnet. The static susceptibility is given by the Langevin formula

$$\chi_n^0 = N\hbar^2 \frac{\gamma_n^2 I(I+1)}{3k_B T},\tag{1.3}$$

where N is the number of nuclei and T the absolute temperature (again we mention that this equation is not valid for solid or liquid ^{3}He at low temperature). The susceptibility being proportional to the square of the gyromagnetic ratio, it remains very small for the nuclear case. A purely static measurement of this quantity is very difficult. (This experiment has been carried out but is of purely historical interest.)

Let us note that eqn (1.3) is valid to an extremely high order of accuracy. If T, γ_n, I, and N are known quantities, this susceptibility is known. This statement seems rather obvious but we have to bear in mind that for an electronic paramagnet the situation is almost always more complicated and in some circumstances an electronic susceptibility can be measured by measuring its ratio to a nuclear susceptibility. This technique will be discussed in another chapter of this book.

1.2. *The resonant technique*

Because of the difficulties encountered in the static detection of nuclear magnetism, experiments use a resonant technique. The principle of this method is to apply an electromagnetic wave of frequency ν to the sample; when this frequency satisfies the relation $h\nu = E_a - E_b$, where E_a and E_b are the energies of two nuclear Zeeman sub-levels, the nuclear system becomes able to extract energy from the electromagnetic wave.

Let us consider a system of N nuclear moments in a static magnetic field H_0 directed along the z-axis. The Hamiltonian of this system is given by the equation

$$\mathcal{H}_0 = -\hbar\gamma_n H_0 \sum_i I_z^i,\tag{1.4}$$

where i is an index that labels the N nuclear spins. The interactions between the nuclear spins are neglected. The energy levels of $\mathcal{H}_0$ are given by the relation

$$E = \sum_i E_{m_i}, \qquad E_{m_i} = \hbar\omega_n m_i,\tag{1.5}$$

where E_{m_i} is the energy for spin i (m_i varies between $+I$ and $-I$), and ω_n is defined as $\omega_n = -\gamma_n H_0$. In the resonant technique the transitions between the levels given by eqn (1.5) are induced by a variable magnetic field directed along the x-axis. We add to the Hamiltonian (1.4) the perturbing term $\mathscr{H}_1$:

$$\mathscr{H}_1 = -\gamma_\mathrm{n} \hbar H_1 \left(\sum_i I_x^i \right) \cos \omega t, \tag{1.6}$$

where H_1 is the amplitude of the variable field. When the frequency $\nu = \omega/2\pi$ is in the vicinity of $\omega_\mathrm{n}/2\pi$ the nuclear system will absorb energy. A straightforward calculation (see Ref. 1, pp. 40 and 41) gives us the absorbed power

$$P \simeq \hbar\omega_\mathrm{n} \frac{\hbar\omega_\mathrm{n}}{k_\mathrm{B} T} \frac{\omega_1^2}{\Delta\omega} N, \tag{1.7}$$

where ω_1 is defined by the relation $\omega_1 = \gamma_\mathrm{n} H_1$ and $\Delta\omega$ is the width of the nuclear levels. This power may be easily estimated but it is perhaps more instructive to compare the result with that obtained in a static measurement.

When a magnetic system is submitted to a variable magnetic field, the absorbed power may be calculated using the classical result

$$P = -\left\langle \mathbf{m} \cdot \frac{\mathrm{d}\mathbf{H}}{\mathrm{d}t} \right\rangle = -\left\langle m_x \frac{\mathrm{d}H_x}{\mathrm{d}t} \right\rangle, \tag{1.8}$$

the average has to be taken over the time variable. This equation shows us that a transverse nuclear moment oscillating at frequency $\omega/2\pi$ must appear in the presence of the oscillating field, otherwise no power would be absorbed. This variable moment will presumably be proportional to H_1 (provided the value of H_1 is not too large), and m_x can be written

$$m_x = H_1\{\chi_\mathrm{n}'(\omega)\cos \omega t + \chi_\mathrm{n}''(\omega)\sin \omega t\}. \tag{1.9}$$

Equation (1.9) is merely a definition of the two functions χ_n' and χ_n'', which are called generalized susceptibilities. Using eqns (1.7)–(1.9) we deduce the value of χ_n'':

$$\chi_\mathrm{n}''(\omega) \simeq \chi_\mathrm{n}^0 \frac{\omega_\mathrm{n}}{\Delta\omega}.$$

As $\Delta\omega$ is usually much smaller than ω_n, the susceptibilities measured in a resonant experiment are greatly enhanced when compared to the static value. This result explains why a resonant method is more sensitive than a static method provided the line width remains small.

2. Influence of the coupling between nuclear moments and other degrees of freedom

Nuclear moments are coupled to electronic degrees of freedom by the hyperfine interaction that we shall discuss in the following chapter.

This interaction produces several observable effects on the nuclear magnetic resonance.

2.1. *Shift of the resonance line*

The quantity that is usually measured is $\chi_n''(\omega)$. It is also possible to measure $\chi_n'(\omega)$, but the knowledge of this other quantity provides no new information because these two functions are coupled by the Kramers–Krönig relations (see Ref. 1, pp. 94–6). As already mentioned, the function $\chi_n''(\omega)$ presents a sharp maximum when ω is in the vicinity of ω_n. For a system of completely free spins in a perfectly homogeneous magnetic field $\chi_n''(\omega)$ would be a delta function:

$$\chi_n''(\omega) \simeq \chi_n^0 \, \delta(\omega-\omega_n).$$

The nuclear moments are not only submitted to the external field H_0 but also to static magnetic fields coming from the electrons. The position of the maximum of χ'' is changed to a value of $\omega = \omega_n + \Delta\omega_n$. The magnitude of the shift $\Delta\omega_n$ varies considerably from case to case. In diamagnetic solids the ratio $\Delta\omega_n/\omega_n$ varies between 10^{-6} and 10^{-4}; in a metal this ratio is larger, varying from 10^{-4} to 10^{-2}. In a paramagnetic substance this shift is a function of the temperature. Finally, in magnetically ordered systems $\Delta\omega_n$ is much larger than ω_n.

2.2. *Structure of the resonance line*

For nuclei with a value of the spin I larger than $\frac{1}{2}$ there are more than two Zeeman sub-levels which need not be equidistant. This fact may be a consequence of the presence of static quadrupolar interactions. In such a case the function χ_n'' presents several sharp peaks.

2.3. *Relaxation time*

Let us consider a system of N nuclear spins $I = \frac{1}{2}$, having the Hamiltonian (1.4). The ratio N_-^0/N_+^0 of the number of spins pointing down to the number of spins pointing up in thermodynamic equilibrium is given by the Boltzmann factor

$$N_-^0/N_+^0 = \exp\!\left(-\frac{\hbar\omega_n}{k_B T}\right),\qquad(1.10)$$

where T is the temperature of the other degrees of freedom. Let us assume that by some method we have produced a situation where the populations are no longer satisfying eqn (1.10). (The simplest method is to apply a large radiofrequency power at the resonance frequency.) Using standard statistical considerations it can be shown that the populations should return to their value in the equilibrium situation.

The mechanism by which the populations equilibrate arises from the coupling between the nuclear system and the other degrees of freedom.

It will be assumed that this coupling can be described by a probability of transition per unit time $W\downarrow$ which gives the probability of transition from the sublevel $m_i = +\frac{1}{2}$ to the sublevel $m_i = -\frac{1}{2}$, and by a probability $W\uparrow$ for the inverse process. In the absence of other external perturbations the rate of change for the populations is given by the equations

$$\frac{dN_+}{dt} = -N_+W\downarrow + N_-W\uparrow,$$

$$\frac{dN_-}{dt} = -N_-W\uparrow + N_+W\downarrow.$$

It is essential to notice that the two probabilities, $W\uparrow$ and $W\downarrow$, are not equal. More precisely, using eqn (1.10) for the equilibrium populations, we deduce that

$$\frac{W\downarrow}{W\uparrow} = \frac{N^0_-}{N^0_+} = \exp\left(-\frac{\hbar\omega_n}{k_B T}\right). \tag{1.11}$$

It may be more convenient to consider the evolution of the nuclear magnetization m_z, which is proportional to the difference between the populations; we obtain, using (1.11),

$$\frac{dm_z}{dt} = -(m_z - m^0_z)\frac{1}{T_1}, \tag{1.12}$$

where T_1 is defined by $\frac{1}{T_1} = W\uparrow + W\downarrow$ and m^0_z is the equilibrium nuclear magnetization. The quantity T_1 is called the longitudinal relaxation time. (The name longitudinal is used because T_1 is the time constant for the evolution of the magnetization along the equilibrium axis.) If the Hamiltonian of the coupling between the nuclear spins and the other degrees of freedom (which, following tradition, we shall call more briefly the 'lattice') is known it is possible to use a perturbation calculation to obtain $W\uparrow$ and $W\downarrow$. At the usual temperatures these two quantities are nearly equal and their ratio given by eqn (1.11) does not differ very much from unity.

2.4. Line width

Now the shape of the resonance (the function χ''_n) will be discussed. This shape will be characterized by the line width $\Delta\omega$ that we have already used for calculating the absorbed power. We first list the possible contributions to $\Delta\omega$.

2.4.1. *Inhomogeneous broadening.* An inhomogeneous broadening is a contribution to $\Delta\omega$ which is due to the fact that spins located at different positions (in space) have different resonance frequencies. The simplest example is provided by a sample in an inhomogeneous magnet. There are other more interesting examples of inhomogeneous broadening; for instance, in a dilute alloy foreign atoms modify the local field in their vicinity.

2.4.2. *Broadening due to coupling with the lattice.* This coupling is at the origin of the relaxation process described in the preceding section. In a rather vague way one may say that the relaxation time T_1 is the lifetime of the nuclear levels: consequently a broadening of the resonance line of the order of $1/T_1$ (in frequency units) is expected. The situation is slightly more subtle: there is a broadening coming from the coupling with the lattice, but this broadening is equal to or larger than $1/T_1$. This result is very important as far as the theory of magnetic resonance is concerned, but for metals a complete discussion would be rather academic as in almost all cases the broadening due to this effect is equal to $1/T_1$. The origin of this possible difference lies in the fact that T_1 is the time constant for the evolution of m_z, whereas the broadening that is the behaviour of the function $\chi_n''(\omega)$ involves, as expected from eqn (1.9), the time constant for the evolution of the transverse magnetization m_x (this time constant is called T_2).

3. Broadening arising from spin–spin interactions

In the previous paragraph the discussion of broadening was very brief, because in most of the cases met in condensed matter the broadening originates from interactions between nuclear moments.

The nuclear moments are always coupled by the classical dipolar coupling. In Chapter 4 it will be shown that in metals other interactions may appear. The energy due to the dipolar interaction between two nuclear moments $\hbar\gamma_1\mathbf{I}_1$ and $\hbar\gamma_2\mathbf{I}_2$ may be written

$$W_{12} = \frac{\gamma_1\gamma_2\hbar^2}{r_{12}^3}\left\{\mathbf{I}_1.\mathbf{I}_2 - 3\,\frac{(\mathbf{I}_1.\mathbf{r}_{12})(\mathbf{I}_2.\mathbf{r}_{12})}{r_{12}^2}\right\}, \qquad (1.13)$$

where $\mathbf{r}_{12}$ is the vector joining the two nuclei.

This interaction produces a broadening of the resonance line of the order of W_{12}/h (in frequency units). The order of magnitude of W_{12} for typical values of the nuclear moments and for a value of r_{12} of the order of a lattice spacing is about 10 kHz. This frequency is still small compared to the nuclear resonance frequency, but is usually larger than that coming from other sources of broadening. The dipolar interaction can produce either a splitting or a broadening of the resonance line. If in a solid two nuclear spins are found separated by a distance much smaller than a lattice spacing (this situation happens very often in hydrated salts for the two hydrogen nuclei of the water molecule), it is convenient to calculate the energy levels for the system of the pair of spins coupled by the dipolar interaction.

The result of the calculation for $I = \frac{1}{2}$ is that the spectrum is divided into two resonance lines, the distance $\delta\omega$ between the lines being given by the equation (Ref. 1, chap. 7)

$$\delta\omega = \frac{3}{4}\frac{\hbar\gamma_n^2}{b^3}(1-3\cos^2\theta),$$

where b is the distance between the nuclei and θ the angle between the direction of the external field and the axis joining the two nuclei. (The two nuclear spins have the same gyromagnetic ratio γ_n.)

In metals it is not usually possible to consider only the energy levels for a pair of nuclear moments and we have to solve the very difficult problem of finding the energy levels for a system of a very large number of nuclear moments with their dipolar interactions. The moments of the resonance line are calculable exactly. These moments are defined as follows:

$$M_2 = \frac{1}{I_n}\int_{-\infty}^{+\infty}(\omega-\omega_n)^2\chi_n''(\omega)\,\mathrm{d}\omega$$

and

$$M_4 = \frac{1}{I_n}\int_{-\infty}^{+\infty}(\omega-\omega_n)^4\chi_n''(\omega)\,\mathrm{d}\omega.$$

I_n is the intensity of the line and is defined as

$$I_n = \int_{-\infty}^{+\infty}\chi_n''(\omega)\,\mathrm{d}\omega.$$

The quantities M_2 and M_4 are called, respectively, the second and fourth moments of the line; ω_n is the resonance frequency in the absence of spin–spin interactions. These moments are known exactly if the interactions are known. For instance, for a system of nuclei coupled by the dipolar interaction, it is found that

$$M_2 = \tfrac{3}{4}\gamma_n^4 I(I+1)\sum_j \frac{(1-3\cos^2\theta_{ij})^2}{r_{ij}^6}, \tag{1.14}$$

where the summation is taken over all the lattice sites occupied by a nuclear spin, r_{ij} is the distance between the nuclei i and j, and θ_{ij} is the angle between the field and the vector $\mathbf{r}_{ij}$. Equation (1.14) shows us that $(M_2)^{\frac{1}{2}}$ is of the order of the frequency shift for the spin i produced by all the nuclear moments. The knowledge of the second moment gives an estimate of the width of the function $\chi''(\omega)$. However this statement is correct only if the shape of $\chi''(\omega)$ is Gaussian (or if the function $\chi''(\omega)$ decreases very fast when $|\omega - \omega_n|$ increases); in some situations the line shape is Lorentzian and the line width is much smaller than $(M_2)^{\frac{1}{2}}$. This problem will be investigated further in Chapter 4.

4. Transient methods in nuclear magnetism (Ref. 1, chap. 3)

So far we have discussed only the so-called steady-state methods where the absorbed power is measured as a function of ω, or for experimental reasons a constant frequency is used and ω_n is changed by changing the external field H_0. For measuring the relaxation times, T_1 and T_2, it is easier to use transient techniques. In these methods the spin system is strongly perturbed by a resonant radiofrequency field during a very short time, and the behaviour of the system as a function of the time is measured after the perturbation is turned off. Among the large variety of transient experiments a very widely used method is that of spin echoes, which permits a measurement of the relaxation times in the presence of a large inhomogeneous broadening.

THE HYPERFINE INTERACTION

I N this chapter the interaction between electrons and a nuclear moment is discussed. It is convenient to separate this interaction into a magnetic and an electric part. We shall consider only the interaction between the nuclear magnetic dipole moment and the electronic magnetic field and the interaction between the nuclear quadrupole moment and the gradient of the electronic electric field. In very accurate experiments using atomic beams a nuclear octupole magnetic moment has been observed, but the magnitude of the energy due to the existence of this moment is extremely small and may be completely neglected in the problem we are concerned with.

1. Magnetic interaction

Let us consider a nuclear moment at the position $\mathbf{R}_i$. The electrons produce at this point a magnetic field $\mathbf{H}_e$ given by (Ref. 1, eqn 6.1):

$$\mathbf{H}_e = -g\beta \sum_e \left[\frac{\mathbf{l}_{ie}}{|\mathbf{r}_e - \mathbf{R}_i|^3} - \frac{\mathbf{s}_e}{|\mathbf{r}_e - \mathbf{R}_i|^3} + 3\, \frac{(\mathbf{r}_e - \mathbf{R}_i)\{\mathbf{s}_e . (\mathbf{r}_e - \mathbf{R}_i)\}}{|\mathbf{r}_e - \mathbf{R}_i|^5} + \right.$$
$$\left. + \frac{8\pi}{3}\, \mathbf{s}_e\, \delta(\mathbf{r}_e - \mathbf{R}_i) \right], \quad (2.1)$$

where $\mathbf{r}_e$ and $\mathbf{s}_e$ are respectively the position and spin of an electron e; $\mathbf{l}_{ie}$ is its orbital moment defined by the equation $\hbar \mathbf{l}_{ie} = (\mathbf{r}_e - \mathbf{R}_i) \wedge \mathbf{p}_e$, where $\mathbf{p}_e$ is the momentum operator ($p_{ez} = -\mathrm{i}(\partial/\partial z_e)$), and so on). The summation is taken over all the electrons. The first term in (2.1) is the orbital hyperfine coupling due to the field produced by the electronic currents. The next two terms describe the dipolar field due to the electronic spin magnetic moment. The last term has no obvious classical interpretation and is called the contact term. This field $\mathbf{H}_e$ interacting with the nuclear moment gives rise to the energy

$$\mathscr{H}_h = -\gamma_n \hbar \mathbf{I}_i . \mathbf{H}_e. \quad (2.2)$$

Let us discuss the influence of the contact part of the hyperfine Hamiltonian. The average value of this operator taken over the electronic variables is non-zero only if the electronic density at the nuclear position is non-zero. For a free atom the electronic density at the nuclear position exists only for s-electrons. In a metal it is possible to expand the one-electron wave functions in spherical harmonics around the nuclear position and only the s-part ($l = 0$) of this expansion will contribute to the average value of this interaction. If the average value of the total hyperfine interaction is taken over the total wave function for an N-electron system, it can be shown that the electrons in complete shells do not contribute to this averaged interaction because their total orbital and spin moments are equal to zero. This result simplifies the evaluation of this interaction but, as will be discussed later, this result is only approximately valid.

For non-transition metals the summation in eqn (2.1) will be performed only over the conduction electrons. For transition metals the inner shells are not complete and the summation has also to be performed over the electrons of the unfilled d- or f-shells. This remark already indicates that the behaviour of a transition metal will differ from that of other metals.

Very often we shall deal with systems having in their ground state a zero average value for the total orbital moment (the orbital moment is quenched), so that the average value of the orbital interaction will vanish.

Further, if the electron cloud has cubic symmetry around the nucleus the average of the dipolar term will also disappear.

The evaluation of the average of the contact term is not easy. As already mentioned, this term exists only if unpaired s-electrons are present. But it is a well-known experimental result that ions or atoms without unpaired s-electrons sometimes show very large scalar hyperfine interactions. An example is the Mn^{++} ion where all but five d-electrons are in filled shells. The cloud of the d-electrons has spherical symmetry (in the Russel–Saunders model the term is $^6S_{5/2}$) so that there is no dipolar and no orbital hyperfine coupling and, as we are considering d-electrons, no average total hyperfine interaction is expected. Experimentally, a large hyperfine coupling is observed.

The explanation of this result is that an electron belonging to the innermost s-shells (like the $1s$- or $2s$-shells) produces an extremely large magnetic hyperfine field and, if the electronic densities are slightly different for the s-electrons with different spin orientations, an important hyperfine interaction will appear. A small difference between the two electronic densities is, in the case of the Mn^{++} ion, produced by the exchange interaction between the inner s-electrons and the unpaired d-electrons. The calculation of the hyperfine field coming from this effect is difficult, for the resultant field so obtained is a small difference between very large fields, but there is no doubt about the validity of this qualitative explanation.[4] This effect, usually called 'core polarization effect', will be particularly important for transition elements where the unpaired electrons have no s-character.

2. Electric interaction

The calculation of electric interaction starts from the equation giving the Coulomb interaction between the electronic and nuclear charge clouds:

$$W = \int\int \frac{\rho_e(r_e)\rho_n(r_n)}{|r_n - r_e|}\, d^3r_e\, d^3r_n, \tag{2.3}$$

where $\rho_e(r_e)$ and $\rho_n(r_n)$ are respectively the electronic and nuclear charge densities. If one excepts the case of the s-electrons, the electronic density is large only at a distance r_e much larger than the nuclear radius and eqn (2.3) may be written

$$W = \int d^3r_n\, \rho(r_n) V(r_n), \tag{2.4}$$

where $V(r_n)$ is the electrostatic potential created by the electrons at the point r_n:

$$V(r_n) = \int d^3r_e\, \frac{\rho(r_e)}{|r_n - r_e|}.$$

The potential $V(r_n)$ varies slowly over the nuclear volume and will be expanded about the centre of the nucleus. There is a difficulty that appears when the electronic density $\rho(r_e)$ exists in the nucleus where the expansion is not obviously valid. We shall neglect this effect which, as shown in Reference 5, p. 517, leads to terms such as the isomeric shift and the isotopic shift which are not important in our case.

$$W = ZV(0) + \sum_\alpha \left.\frac{\partial V}{\partial\alpha}\right|_{r=0} \times \int \alpha\rho(r_n)\, d^3r_n +$$
$$+ \tfrac{1}{2}\sum_{\alpha,\beta} \left.\frac{\partial^2 V}{\partial\alpha\partial\beta}\right|_{r=0} \times \int \alpha\beta\rho(r_n)\, d^3r_n + ..., \tag{2.5}$$

where α or β denotes one of the coordinates x, y, or z and Z is the total nuclear charge.

The second term of the expansion vanishes because it is proportional to the electric nuclear dipole moment. Let us estimate the ratio of the third term to the first one. (This evaluation will also justify the neglect of the higher-order terms in the expansion (2.5) since the next term involves the fourth-order derivatives of V and is smaller by the square of this ratio.) The potential V varies appreciably over distances of the order of the Bohr radius a_0; the second derivative of V is roughly equal to $V(0)/a_0^2$, and the ratio of the two terms is of the order of $(R/a_0)^2$, where R is the nuclear radius. Using the numerical values $R = 10^{-12}$ cm and $a_0 = 10^{-8}$ cm, the third term has a magnitude of 10^{-7} eV or 20 MHz (the next neglected term will be smaller than 1 Hz). The third term in (2.5) is called the quadrupolar hyperfine interaction.

The part of this term involving the nuclear variables can be transformed using the Wigner–Eckart theorem, and the quadrupolar interaction may be written

$$W_Q = \frac{eQ}{6I(2I-1)} \sum_{\alpha,\beta} \frac{\partial^2 V}{\partial\alpha\partial\beta}\bigg|_{r=0} \{\tfrac{3}{2}(I_\alpha I_\beta + I_\beta I_\alpha) - \delta_{\alpha\beta} I(I+1)\}. \quad (2.6)$$

Q is a quantity characteristic of a given nucleus and is called the electric quadrupole moment.

This interaction exists only for nuclei with spin larger than one-half. Equation (2.6) may be written in a simpler form by another choice of the coordinate axes (X, Y, Z); we note that the tensor $\partial^2 V/\partial\alpha\partial\beta$ is symmetric and by a change of the frame of reference it can be diagonalized. This leaves three parameters $\partial^2 V/\partial X^2$, $\partial^2 V/\partial Y^2$, $\partial^2 V/\partial Z^2$ to characterize the electronic field, but by using the Laplace equation we can eliminate one of them, so that the interaction may now be written as

$$W_Q = \frac{e^2 qQ}{4I(2I-1)}\left\{3I_z^2 - I(I+1) + \frac{\eta}{2}(I_+^2 + I_-^2)\right\}, \quad (2.7)$$

with the following definitions:

$$eq = \frac{\partial^2 V}{\partial Z^2}, \qquad \eta = \left(\frac{\partial^2 V}{\partial X^2} - \frac{\partial^2 V}{\partial Y^2}\right)\bigg/\left(\frac{\partial^2 V}{\partial Z^2}\right),$$

and $I_\pm = I_x \pm iI_y$. η is called the asymmetry parameter.

The quadrupolar coupling may also be expressed using spherical instead of cartesian coordinates. This method will be used in Chapter 5. The gradients of the electric field are in this case simple combinations of the second-order spherical harmonics of the electron coordinates and again by using the Wigner–Eckart theorem these harmonics are written as functions of the orbital moment L (or J). This form of W_Q is convenient when working with free atoms (or free ions) because the system is invariant by rotation and L and J are good quantum numbers; but in metals, where the symmetry is lower, the form given by eqn (2.7) is more convenient.

As for the magnetic term, we note that the electrons in complete shells do not contribute to the field gradients because their charge cloud is isotropic.

In metals two contributions to the electric field gradients are found: one from the conduction electron charge cloud and the other from the ionic charges. (For transition metals there is also a contribution from the electrons in the incomplete inner shell.)

But for the quadrupolar interaction a serious complication also arises due to the distortion of the charge cloud of the complete inner shells. Let us first consider the electric field due to the ions. The interaction of this field with the originally spherical inner electron cloud causes it to be distorted and so to produce an electric field gradient.

Let us call one of the components of the gradient tensor produced by the external charges $V^e_{\alpha\beta}$ and a gradient produced by an inner shell electron $V^i_{\alpha\beta}$. If ψ_0 is the electronic ground-state wave function in the absence of external charges, the following relation is obeyed:

$$\langle\psi_0|V^i_{\alpha\beta}|\psi_0\rangle = 0.$$

Adding the external charges changes the wave function, which becomes $\psi_0+\delta\psi_0$. The average electric field gradient is given by the equation

$$V_{\alpha\beta} = \langle\psi_0+\delta\psi_0|V^e_{\alpha\beta}+V^i_{\alpha\beta}|\psi_0+\delta\psi_0\rangle$$
$$\simeq \langle\psi_0|V^e_{\alpha\beta}|\psi_0\rangle+\langle\delta\psi_0|V^i_{\alpha\beta}|\psi_0\rangle+\langle\psi_0|V^i_{\alpha\beta}|\delta\psi_0\rangle,$$

where only the lowest order terms are taken into account. The second and third terms have no reason to be small compared to the first one. Even if the change in the wave function $\delta\psi_0$ is small, the gradients $V^i_{\alpha\beta}$ are much larger than $V^e_{\alpha\beta}$ because the electric field created by an inner electron is much larger than the fields due to the ionic charges.

The calculation of $V_{\alpha\beta}$ gives the following result. The quantities V^e are replaced by $V^e(1+\gamma)$, where γ is called the anti-shielding factor. This quantity depends on the atom we are considering. γ is small and negative for light atoms and becomes large and positive for heavy atoms; for instance, $\gamma \simeq 4$ for Na^+ and $\gamma \simeq 70$ for Rb^+. The calculation of the electric field gradient due to the conduction electrons is a different problem; it is no longer possible to consider the field as produced by external point charges since the conduction electron charge cloud overlaps the inner electronic shells. There is still a correction due to the distortion of the inner shells but the magnitude of this correction is much smaller. The total electric field gradient is

$$V_{\text{total}} = V_e(1+\gamma)+(1-R)V_c,$$

where V_c is the uncorrected gradient due to the conduction electrons and R is a correcting factor that is typically smaller than 0.2.

The reason why the factor γ is sometimes very large whereas R remains small can be understood by using a very crude estimate. Let us assume that the inner shell is a spherical shell of electrons that are at distance R_1 from the nucleus. An external charge at a distance R_{ex} much larger than R_1 produces a gradient that varies as $1/R_{ex}^3$ and a small distortion of the shell brings up a large contribution varying as $1/R_1^3$. On the other hand, if the charge is at a distance R_{int} with $R_{int} < R_1$ there is still a distortion but the largest contribution will be due to the $1/R_{int}^3$ term.

Because the correcting factor γ is large in many ions, the term due to the conduction electrons will be usually less important and the quadrupolar effects observed in pure metals are very similar to the effects encountered in ionic crystals. There are two exceptions to this rule: first for transition metals, where the d- (or f-) electrons may contribute to the quadrupolar interaction, and secondly in alloys, where the change in the electronic density around the impurity gives large quadrupolar effects.

The separation into two parts of the electric field gradient in metals is not the only possible approach. For calculating the gradient by this method a knowledge of the electronic wave function is required. In some cases, such as the liquid metals, the wave functions are not known and it is more convenient here to consider the electrons as screening the ionic charges and to evaluate the electric field gradient using a screened ionic potential. This method will be discussed in Chapter 8 when the nuclear resonance in liquid metals will be considered.

3. Some considerations of the relative order of magnitude of the two parts of the hyperfine interaction

If the two interactions are calculated for a given unpaired electron, it is found that they are of the same order of magnitude for all nuclei and that they both vary between 1 and 1000 MHz. The magnetic interaction (neglecting the orbital term) is a function of the spin variables, while the quadrupolar interaction depends only on the orbital variables (more precisely this interaction can be expressed in terms of second-order tensors of the orbital moment). In the systems we are considering the ground state usually has a spin degeneracy but no orbital degeneracy and the effects produced by the two interactions are very different. The quadrupolar coupling has a very well-defined value given by calculating the average value of the orbital tensors in the orbitally non-degenerate ground state, whereas for the magnetic interaction we have to take the average over the various degenerate spin states. The averaged magnetic interaction is thus greatly reduced. An exception to this rule occurs for magnetically ordered systems where the magnetic interaction may reach its maximum value.

3

SOME PROPERTIES OF THE
METALLIC STATE

IN this chapter we shall not try to give a complete review of the properties of metals. Our aim is only to define the notations we shall need and to discuss the usual approximations made for describing the metallic properties. The discussion will consider in more depth the properties that can be studied using magnetic resonance methods. Certain very important features of the metallic state, such as transport and optical properties, will be discussed extremely briefly or not at all. The complete description of a metal, even in the simplest cases, is an extremely complicated problem involving the description of the motion of about 10^{22} electrons and nuclei. The first approximation—usually a very good one—consists of a separation of the electronic and nuclear motions. The problem then is to describe the motion of the electrons in the regular lattice of the nuclear charges assumed to be at rest.

There are various ways of describing this motion, none of them perfect. The choice of a given model will depend both on the metal we are considering and on the property we are looking at. The first simple model consists in starting from the behaviour of a free-electron gas. The problem of a free-electron gas without interaction between the electrons can be exactly solved, and with some complications it is possible to take into account, at least partially, the interactions between electrons. This model is not too unrealistic for understanding some of the properties of metals like the alkali metals that have only one energy band. In these metals the electronic energy comes mainly from the kinetic energy of the electrons, and the presence of a periodic potential energy due to the ionic charges can sometimes be neglected in first approximation. There are various models that are improvements of this simple model; they all start from wave functions involving plane waves (Ref. 7, chap. 13). A different approach is the strong binding model. Here the starting wave functions are atomic wave functions, fitting the model to the study of metals with narrow energy bands.

1. The free-electron gas: the Landau quasi-particle model

Let us first consider the properties of a gas of N electrons without interaction. The wave functions are plane waves. The energy levels and wave functions are characterized by a wave vector $\mathbf{k}$, the energy is given by the relation

$$E_{\mathbf{k}} = \frac{\hbar^2}{2m} |\mathbf{k}|^2, \tag{3.1}$$

and the wave function

$$\phi_{\mathbf{k}}(\mathbf{r}) = \mathrm{e}^{\mathrm{i}\mathbf{k}\cdot\mathbf{r}}. \tag{3.2}$$

(This wave function is normalized in the unit volume.)

Another degree of freedom that has to be taken into account is the direction of the electronic spin. If there is no external magnetic field the energies corresponding to electrons with the same wave vector but with two different spin orientations are the same. The wave function will be written $\phi_{\mathbf{k},\sigma}$ or $\phi_{\mathbf{k}\uparrow}$ and $\phi_{\mathbf{k}\downarrow}$, where the arrows indicate the spin orientation with respect to a given axis.

The total wave functions of the system Ψ, are Slater determinants using N one-electron wave functions. The ground-state wave function Ψ_0 is obtained by using the N wave functions $\phi_{\mathbf{k},\sigma}$ corresponding to the lowest energy levels. The maximum value for the energy of an occupied state in the ground state is called the Fermi energy and will be written E_{F}. The Fermi wave vectors $\mathbf{k}_{\mathrm{F}}$ are defined by the equation

$$E_{\mathrm{F}} = \frac{\hbar^2}{2m} |\mathbf{k}_{\mathrm{F}}|^2. \tag{3.3}$$

For a free-electron gas E_{F} and $|\mathbf{k}_{\mathrm{F}}|$ are known as functions of N/V, where N is the number of electrons and V the volume occupied by the electron gas. The wave vectors $\mathbf{k}$ of all the occupied one-electron states have amplitude $|\mathbf{k}| < |\mathbf{k}_{\mathrm{F}}|$ in the ground state of the system. The Fermi surface is the surface in the reciprocal space described by the extremities of the Fermi wave vectors; in this simple model this surface as shown by eqn (3.3) is a sphere. An important quantity that appears very often in the calculations, the density of states, is defined as follows: $g(E)\,\mathrm{d}E$ is the number of one-electron energy levels having an energy between E and $E+\mathrm{d}E$. In this simple isotropic model for a sample of volume unity the following value is found:

$$g(E) = \frac{1}{2\pi^2} \left(\frac{2m}{\hbar^2}\right)^{3/2} E^{1/2}. \tag{3.4}$$

Finally, the value of the probability $n(E_{\mathbf{k}})$ of finding a state of energy $E_\mathbf{k}$ occupied at a given temperature T is given by the Fermi–Dirac function

$$n(E_\mathbf{k}) = \left\{\exp\left(\frac{E_\mathbf{k} - E_\mathrm{F}}{k_\mathrm{B} T}\right) + 1\right\}^{-1}. \tag{3.5}$$

For usual electronic densities found in metals, E_F is a rather large energy; we find for the Fermi temperature T_F ($k_\mathrm{B} T_\mathrm{F} = E_\mathrm{F}$) typical values of the order of 10 000 K; and a good approximation is always obtained by assuming that $k_\mathrm{B} T$ is much smaller than E_F. (This statement will not be valid for semiconductors or semimetals where T_F has a small value.) With this approximation the function $n(E)$ differs from the values 0 or 1 only if E is in a region of width $k_\mathrm{B} T$ around the Fermi energy. This is very useful for estimating the order of magnitude of the various effects encountered in metals.

For calculating the matrix element of a given operator between two N-electron wave functions, it is very often more convenient to use the second quantization formalism (see Ref. 7, p. 76 and Ref. 8, p. 221). Only the notations and the main properties of this formalism will be presented.

We define a creation operator written $C_{\mathbf{k},\sigma}^{+}$ which has the following properties. If $C_{\mathbf{k},\sigma}^{+}$ is applied to a P-electron wave function, a $P+1$ electron wave function is obtained, the last electron being in the state $\mathbf{k},\sigma$. We also define a destruction operator $C_{\mathbf{k},\sigma}$ which destroys the electron in the state $\mathbf{k},\sigma$.

Thus starting from Φ_0, the vacuum wave function, and applying N creation operators, we can build all the N-electron wave functions. The very important result is that the required antisymmetry of the N-electron wave functions is obtained automatically if the operators obey the following anticommutation rules:

$$\left. \begin{aligned} C_{\mathbf{k},\sigma}^{+} C_{\mathbf{k}',\sigma'} + C_{\mathbf{k}',\sigma'} C_{\mathbf{k},\sigma}^{+} &= \delta_{\mathbf{k}\mathbf{k}'} \delta_{\sigma\sigma'} \\ C_{\mathbf{k},\sigma}^{+} C_{\mathbf{k}',\sigma'}^{+} + C_{\mathbf{k}',\sigma'}^{+} C_{\mathbf{k},\sigma}^{+} &= 0 \\ C_{\mathbf{k},\sigma} C_{\mathbf{k}',\sigma'} + C_{\mathbf{k}',\sigma'} C_{\mathbf{k},\sigma} &= 0 \end{aligned} \right\}. \tag{3.6}$$

Let us consider an operator A which can be written as $A = \sum_\mathrm{e} a(\mathbf{r}_\mathrm{e}, s_\mathrm{e})$ where $a(\mathbf{r}_\mathrm{e}, s_\mathrm{e})$ is only a function of the coordinates (and spin) of the electron e. It can be shown that all the properties of the matrix elements of A between two N-electron wave functions are reproduced if A is written as

$$A = \sum_{\substack{\mathbf{k},\sigma \\ \mathbf{k}',\sigma'}} a(\mathbf{k}, \sigma; \mathbf{k}', \sigma') C_{\mathbf{k},\sigma}^{+} C_{\mathbf{k}',\sigma'}, \tag{3.7}$$

with the definition

$$a(\mathbf{k}, \sigma; \mathbf{k}', \sigma') = \langle \phi_{\mathbf{k}\sigma} | a(\mathbf{r}, s) | \phi_{\mathbf{k}'\sigma'} \rangle. \tag{3.8}$$

For free electrons the spatial part of $\phi_{\mathbf{k},\sigma}$ is given by eqn (3.2), but eqns (3.7) and (3.8) are valid for any complete set of one-electron wave functions $\phi_{\mathbf{k},\sigma}$.

It is also possible to write as a function of creation and destruction operators the expression of an operator that is a function of the co-ordinates of two electrons. An equation similar to eqn (3.7) is obtained but now four operators are involved instead of two. Finally, the thermal average of a product of two operators has the value

$$\langle C^{+}_{\mathbf{k},\sigma} C_{\mathbf{k}',\sigma'} \rangle = n(E_{\mathbf{k},\sigma})\delta_{\mathbf{k}\mathbf{k}'}\delta_{\sigma\sigma'}, \tag{3.9}$$

where $n(E)$ is as defined by eqn (3.5). The averages of products of two creation operators (or two destruction operators) are equal to zero.

The next step is to introduce the interaction between electrons. We shall explain only a semi-empirical approach due to Landau (more details are found in Refs. 9 and 10). This model was originally built to study the low-temperature properties of liquid helium-3 (which is an assembly of uncharged fermions, and where no lattice is present) but the basic ideas are useful for understanding the properties of an interacting electron gas in a physical sense. The model has been very useful for interpreting recent experiments on the resonance of conduction electrons, where the results cannot be explained if the interactions between electrons are neglected. Our description here is short but more details will be given in Chapter 10, where the electronic resonance of conduction electrons is discussed.

The total energy of the system is assumed to be a function of the occupation numbers $n_{\mathbf{k}}$. For a non-interacting system this statement is rather obvious, the total energy being simply given by the equation

$$E = \sum_{\mathbf{k}} E_{\mathbf{k}} n_{\mathbf{k}}.$$

This relation will no longer be assumed. The model is built for studying only the low energy excitations of the system, such as creating an electron with a wave vector slightly larger than the Fermi wave vector, while destroying an electron with a wave vector slightly smaller than k_{F}. In a non interacting system if the occupation numbers $n_{\mathbf{k}}$ change by a small amount $\delta n_{\mathbf{k}}$ the total energy is changed by the quantity δE:

$$\delta E = \sum_{\mathbf{k}} \delta n_{\mathbf{k}} E_{\mathbf{k}}.$$

The idea of Landau was to take this relation as a zero-order approximation and to expand δE as a function of $\delta n_{\mathbf{k}}$:

$$\delta E = \sum_{\mathbf{k}} \epsilon^{0}_{\mathbf{k}} \delta n_{\mathbf{k}} + \tfrac{1}{2}\sum_{\mathbf{k},\mathbf{k}'} f(\mathbf{k},\mathbf{k}')\delta n_{\mathbf{k}} \delta n_{\mathbf{k}'}. \tag{3.10}$$

The interactions are taken into account by the presence of the function $f(\mathbf{k}, \mathbf{k}')$ and also by a possible difference between $\epsilon_\mathbf{k}^0$ of (3.10) and the $E_\mathbf{k}$ of a free-electron gas.

Using eqn (3.10) we find that if the occupation number $n_\mathbf{k}$ changes by $\delta n_\mathbf{k}$, the change in energy we shall call $\epsilon_\mathbf{k}$ has the value

$$\epsilon_\mathbf{k} = \epsilon_\mathbf{k}^0 + \sum_\mathbf{k} f(\mathbf{k}, \mathbf{k}')\delta n_{\mathbf{k}'}. \tag{3.11}$$

$\epsilon_\mathbf{k}$ is the energy of an excitation of wave vector $\mathbf{k}$ in the presence of other excitations. This excitation is called a quasi-particle and $\epsilon_\mathbf{k}$ is its energy. The theory is restricted to the study of excitations $\delta n_\mathbf{k}$ and $\delta n_{\mathbf{k}'}$ with $\mathbf{k}$ and $\mathbf{k}'$ in the vicinity of the Fermi wave vector and the function f is only a function of the angle $\theta_{\mathbf{k}\mathbf{k}'}$ between the two wave vectors. Again the spin variables are introduced by defining the quantities $\delta n_{\mathbf{k},\sigma}$ and $\epsilon_{\mathbf{k},\sigma}$ and a function $f(\mathbf{k}\sigma, \mathbf{k}'\sigma')$. If there is no applied magnetic field the system must be invariant to any rotation in the spin space, thus $\epsilon_{\mathbf{k}\uparrow}^0 = \epsilon_{\mathbf{k}\downarrow}^0$ and f is only a function of the relative orientation of the two spins. The interaction can be written as

$$f(\mathbf{k}\sigma, \mathbf{k}'\sigma') = f_0'(\mathbf{k}, \mathbf{k}') + \boldsymbol{\sigma}.\boldsymbol{\sigma}' f_e'(\mathbf{k}, \mathbf{k}'),$$

or in the following equivalent form:

$$f(\mathbf{k}\sigma, \mathbf{k}'\sigma') = f_0(\mathbf{k}, \mathbf{k}') + \delta_{\sigma\sigma'} f_e(\mathbf{k}, \mathbf{k}'), \tag{3.12}$$

where in this isotropic model f_0 and f_e (or f_0' and f_e') are simply functions of $\theta_{\mathbf{k}\mathbf{k}'}$.

The relation between the known coulomb interaction for two electrons and the functions f is by no means simple. The functions f describe the interactions between two quasi-particles $\mathbf{k}$ and $\mathbf{k}'$ when all the other electrons have their energies smaller than E_F. This model can be justified by starting from the complete Hamiltonian of the system and using the standard techniques of the many-body problem.

2. The electron in a periodic lattice

In a metal the electrons are moving in the periodic potential coming from the nuclear charges surrounded by the inner-shell electrons. (We assume that the effect of the nuclei and inner-shell electrons can be treated as a potential.) The one-electron energy levels are obtained by solving the Schrödinger equation, taking this periodic potential into account. A general theorem tells us that in a periodic potential the wave function must have the form

$$\phi_\mathbf{k}(\mathbf{r}) = U_\mathbf{k}(\mathbf{r})e^{i\mathbf{k}.\mathbf{r}}, \tag{3.13}$$

where the function $U_\mathbf{k}(\mathbf{r})$ has the periodicity of the lattice. The function $U_\mathbf{k}$, and consequently $\phi_\mathbf{k}$, are usually normalized in the volume V of the sample, but it is sometimes convenient to use a function $\psi_\mathbf{k}$ normalized in the atomic volume Ω; $\psi_\mathbf{k}$ is then given by the relation

$$\psi_\mathbf{k}(\mathbf{r}) = \sqrt{\left(\frac{V}{\Omega}\right)}\phi_\mathbf{k}(\mathbf{r}). \tag{3.14}$$

It is also known that it is no longer necessary to consider all the possible values for the vectors $\mathbf{k}$, and only the vectors $\mathbf{k}$ inside the first Brillouin

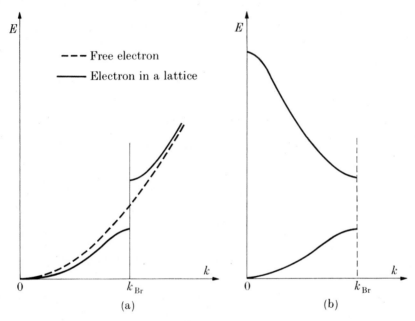

FIG. 3.1. Energy of the electron as a function of the wave vector. In (b) the energy with the wave vector restricted to the first Brillouin zone is shown.

zone will be considered. If this restriction is made, there are several wave functions and energy levels corresponding to a given wave vector $\mathbf{k}$ (see Fig. 3.1). A wave function will be written as

$$\phi_{\mathbf{k},n} = U_{\mathbf{k},n}\,e^{i\mathbf{k}.\mathbf{r}}, \tag{3.15}$$

where the index n labels the various energy bands.

If the wave vector crosses the boundary of the Brillouin zone the energy changes in a discontinuous way. As before, the Fermi energy is defined as the maximum value of the energy of the occupied lowest levels. The Fermi surface is still defined by the equation $E(\mathbf{k}) = E_\mathrm{F}$, but this surface has no reason to be a sphere and, in general, the Fermi surface possesses the symmetry of the reciprocal lattice space. In some metals the Fermi energy crosses only one energy band but more complicated situations are encountered.

As already mentioned, the second quantization formalism may be used and eqn (3.7) is valid provided the set of functions $\phi_{\mathbf{k},n,\sigma}$ is complete. Finally, the Landau model may be generalized to describe the behaviour of interacting electrons in the metal. However, as the system is no longer isotropic this model loses its simplicity. As an example, $\epsilon_{\mathbf{k}}^{0}$ is now a function of the orientation of the vector $\mathbf{k}$ and the function $f(\mathbf{k},\mathbf{k}')$ is not simply a function of the angle between the wave vectors. Nevertheless the model is still useful if the Fermi surface is not too different from a sphere and again if one considers only wave vectors in the vicinity of the Fermi surface.

3. The response of a metal to a magnetic perturbation

As we already noticed in the previous chapter, one of the more important terms in the hyperfine coupling is the magnetic contact interaction. This term may be interpreted as a magnetic field, varying in space, applied to the electronic spin system. Therefore many of the nuclear magnetic resonance properties in metals are known if the response of the electronic spin system to an external magnetic perturbation varying in space (and eventually in time) is known. The response of the electronic system to a field varying in space is described by the generalized susceptibilities. The knowledge of these quantities is also very useful for understanding the electronic resonance properties; it will be shown in the last chapter that an experiment of electronic resonance on conduction electrons is merely a measurement of a generalized susceptibility.

3.1. *Static-spin susceptibility*

The metal is submitted to a static uniform magnetic field H_0. If M_S is the part of the electronic magnetization due to the electronic spin induced by this field, the static susceptibility χ_S is defined by the relation

$$M_S = \chi_S H_0. \tag{3.16}$$

The calculation of χ_S for a gas of N free electrons is as follows. In the presence of the magnetic field the energies of the electrons with different spin orientations (but with the same $\mathbf{k}$) differ by the amount

$$\Delta E = \hbar\omega_0 = g\beta H_0.$$

The electronic ground state is again built by occupying the lowest energy levels and consequently there are more electrons with their spin pointing down than electrons with a spin pointing up (see Fig. 3.2). The spin magnetization is given by the relation

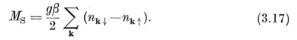

$$M_S = \frac{g\beta}{2} \sum_{\mathbf{k}} (n_{\mathbf{k}\downarrow} - n_{\mathbf{k}\uparrow}). \tag{3.17}$$

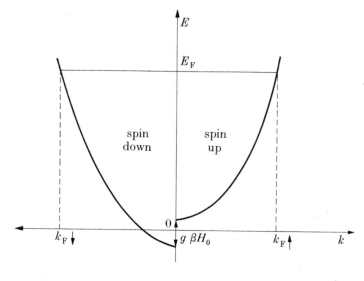

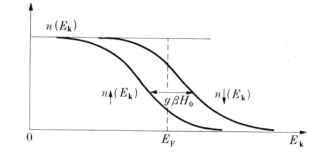

FIG. 3.2. The free-electron gas in an external magnetic field.

Instead of summing over the wave vectors, we perform an integration over the energies:

$$M_S = g\beta \frac{V}{2} \int_0^\infty dE_{\mathbf{k}} \frac{g(E_{\mathbf{k}})}{2} \{n_\downarrow(E_{\mathbf{k}}) - n_\uparrow(E_{\mathbf{k}})\}, \tag{3.18}$$

where $E_{\mathbf{k}}$ is the energy neglecting the electronic Zeeman energy. We use the density-of-state function $g(E_{\mathbf{k}})V/2$ because we consider the density of states for a given direction of the spin, whereas the value of $g(E)$ given by eqn (3.4) is defined for the two possible spin orientations (V is the volume). As the electronic Zeeman energy is small, the difference between the occupation numbers may be written

$$n_{\downarrow}(E_{\mathbf{k}}) - n_{\uparrow}(E_{\mathbf{k}}) = \frac{\partial n(E_{\mathbf{k}})}{\partial E_{\mathbf{k}}} \Delta E = -g\beta H_0 \frac{\partial n(E_{\mathbf{k}})}{\partial E_{\mathbf{k}}},$$

and at low temperatures the function $\partial n/\partial E$ may be replaced by the delta function $-\delta(E_{\mathbf{k}} - E_{\mathrm{F}})$ (see Fig. 3.2). We get

$$\chi_{\mathrm{S}} = \frac{V}{4}\,(g\beta)^2 g(E_{\mathrm{F}}). \tag{3.19}$$

Even for free electrons there are two complications appearing in the calculation of the static susceptibility χ_{S}:

(1) In the calculation the electronic wave function was assumed to be a plane wave. It is well known (Ref. 8, p. 474) that the wave function of a charged free particle in a static magnetic field is not at all correctly described by a plane wave. The energy levels and the density of states are completely modified. The density of states function now diverges to infinity each time the energy reaches a value of the form

$$E = \hbar\omega_{\mathrm{c}}(n + \tfrac{1}{2}),$$

where ω_{c} is the cyclotron resonance frequency and n an integer. These oscillations induce an oscillatory behaviour for various properties of metals, the best example being the oscillations observed in the variation of the diamagnetic susceptibility, called the De Haas–Van Alphen effect. Looking at eqn (3.19) one may suspect that large oscillations will also appear in the variation of χ_{S} as a function of the external field. However eqn (3.19) is valid only if the density of state function $g(E)$ is a slowly varying function of E, varying slowly compared to $k_{\mathrm{B}}T$ and $\hbar\omega_0$; if these conditions are not fulfilled, it is necessary to use the exact equation (3.18). Knowing the variation of the function $g(E)$ with the field and the occupation numbers the calculation of χ_{S} can be done exactly. The calculation is tedious and the following result is obtained. The oscillations of $g(E)$ are strongly reduced by the integration; a small oscillation persists in χ_{S} and its amplitude compared to the average value of χ_{S} is given by the number

$$\frac{k_{\mathrm{B}}T}{\hbar\omega_0}\left(\frac{\hbar\omega_{\mathrm{c}}}{2E_{\mathrm{F}}}\right)^{\frac{1}{2}} \exp\left(-\frac{2\pi^2 k_{\mathrm{B}}T}{\hbar\omega_{\mathrm{c}}}\right),$$

which is almost always a small number.[11]

(2) In defining the density of states we have also implicitly assumed that a large number of energy levels are present in the vicinity of the Fermi energy. For very small particles of metals this condition is no longer fulfilled. The static susceptibility will differ from the value given in our former description when the distance Δ between the energy levels becomes larger than $k_B T$. The distance Δ is of the order of E_F/N. When Δ is large compared to $k_B T$ one must consider separately the behaviour of particles where N is odd or even. For even values of N, χ_S decreases as the temperature decreases and when T is below the critical value $k_B T_c = \Delta$ the system tends to become diamagnetic. On the other hand, when N is an odd number, χ_S increases below T_c and varies like $1/T$ (provided $g\beta H_0 \ll k_B T$).[12,13] Let us estimate the size of a particle giving for Δ/k_B a value of 1 K, if $T_F = 10^4$ K, we find $N = 10^4$ atoms for a lattice spacing of 3 Å. This figure corresponds approximately to a sphere of 100 Å diameter.

Let us calculate χ_S using the Landau model. We write eqn (3.17) in the form

$$\chi_S = M_S/H_0 = \frac{g\beta}{2} \sum_k (\delta n_{k\downarrow} - \delta n_{k\uparrow})/H_0, \qquad (3.20)$$

where $\delta n_{k\uparrow}$ and $\delta n_{k\downarrow}$ are the changes in the occupation numbers produced by the presence of the magnetic field. The energy of an electron with a given wave vector depends on its spin orientation, and the Fermi wave vectors corresponding to the two spin orientations differ by a small amount $\Delta k_F = k_{F\downarrow} - k_{F\uparrow}$. The susceptibility is proportional to Δk_F, which can be calculated if the change in the quasi-particle energy for the two different spin orientations $\Delta \epsilon$ is known. These two quantities are related by the equation

$$\Delta k_F = \Delta \epsilon \Big/ \left(\frac{d\epsilon}{dk}\right)_{k=k_F}.$$

(In this section we write k instead of the more correct notation $|k|$.)

Let us calculate $\Delta \epsilon$. There are two contributions to this change of energy, one from ϵ_k^0, which is the electronic Zeeman energy, the other from the interaction energy because the occupation numbers are changed by the presence of a magnetic field. We start from the two equations

$$\epsilon_{k\uparrow} = \epsilon_k + \tfrac{1}{2}g\beta H_0 + \sum_{k',\sigma'} f(k\uparrow, k'\sigma')\delta n_{k'\sigma'},$$

$$\epsilon_{k\downarrow} = \epsilon_k - \tfrac{1}{2}g\beta H_0 + \sum_{k',\sigma'} f(k\downarrow, k'\sigma')\delta n_{k'\sigma'}.$$

By subtracting these two equations $\Delta\epsilon$ is obtained:

$$\Delta\epsilon(\mathbf{k}) = g\beta H_0 + \sum_{\mathbf{k}'} f_e(\mathbf{k}, \mathbf{k}')(\delta n_{\mathbf{k}'\uparrow} - \delta n_{\mathbf{k}'\downarrow}).$$

All the wave vectors are at the Fermi surface. This quantity in this isotropic model does not depend on the direction of $\mathbf{k}$, and therefore it is possible to take the average value of the right-hand side of the equation over the direction of $\mathbf{k}$. The equation now becomes

$$\Delta\epsilon = g\beta H_0 + \sum_{\mathbf{k}'} (\delta n_{\mathbf{k}'\uparrow} - \delta n_{\mathbf{k}'\downarrow})\langle f_e(\mathbf{k}, \mathbf{k}')\rangle,$$

where the average is taken over all the relative orientations of $\mathbf{k}$ and $\mathbf{k}'$. As the quantity $\delta n_{\mathbf{k}'\uparrow} - \delta n_{\mathbf{k}'\downarrow}$ is proportional to the field H_0, this equation can be written

$$\Delta\epsilon = g\beta H_0(1+\alpha),$$

with

$$g\beta H_0\,\alpha = \sum_{\mathbf{k}} (\delta n_{\mathbf{k}\uparrow} - \delta n_{\mathbf{k}\downarrow})\langle f_e(\mathbf{k}, \mathbf{k}')\rangle.$$

Using eqn (3.20) α can be expressed as a function of χ_S,

$$\alpha = -\frac{2\chi_S}{(g\beta)^2}\,\langle f_e(\mathbf{k}, \mathbf{k}')\rangle. \tag{3.21}$$

On the other hand, as the susceptibility is proportional to $\Delta\epsilon$, the following relation may be written:

$$\chi_S = \chi_S^0(1+\alpha), \tag{3.22}$$

where χ_S^0 is the spin susceptibility neglecting the term f_e in the interaction. Using eqns (3.21) and (3.22), we obtain

$$\chi_S = \frac{\chi_S^0}{1+B_0}, \tag{3.23}$$

with

$$B_0 = \frac{2\chi_S^0}{(g\beta)^2}\,\langle f_e(\mathbf{k}, \mathbf{k}')\rangle.$$

In this equation χ_S^0 was defined as the susceptibility in the absence of the exchange term, but is not exactly the susceptibility χ_S^F in the absence of interaction. The interactions produce another correction that appears because χ_S is proportional to $(d\epsilon/dk)_{k_F}^{-1}$ and this quantity is modified by the interactions. Traditionally, this second correction is expressed as a change in the electronic mass $[(1/\hbar)(d\epsilon/dk)$ is equal to the velocity $v_k = p/m$, where p is the momentum], and the equation is written

$$\chi_S = \frac{m^*}{m}\,\frac{\chi_S^F}{1+B_0}, \tag{3.24}$$

where m^* is the corrected electronic mass. The ratio m^*/m in this model is a function of f_0; but we shall not use this expression because in metals this quantity is also changed by other interactions such as the electron–phonon interactions and also by the band structure.

There are other more involved calculations of the electronic susceptibility.[14,15]

Finally, let us mention another complication. In a real metal the electronic spins are coupled to the orbital moment by the spin-orbit coupling interaction. This interaction is particularly large for heavy elements and has the effect of making the spatial parts of the wave function different for the two spin orientations. As the spin-orbit coupling interaction is also periodic and if one assumes that the orbital moment is quenched, this interaction has non-diagonal elements only between wave functions with the same wave vector **k** but belonging to different energy bands. The magnitude of the perturbation due to the spin-orbit coupling will be important when the levels of two bands (for a given value of **k**) occur at a comparable energy. One of the effects of this coupling is to change the electronic Zeeman energy, a question that will be discussed in Chapter 10.

A purely static measurement of the spin susceptibility in a metal is difficult because there are several other contributions to the induced magnetic moment such as the diamagnetic magnetization of the conduction electrons and the ionic diamagnetic magnetization. All these contributions are very often of comparable magnitude. It will be shown later that χ_S has been measured by a resonant technique in a small number of metals.

3.2. *Time-dependent susceptibilities*

This concept has already been introduced in the first chapter for the nuclear spin system. The electronic spin system is submitted to a time-oscillating but spatially uniform field directed along the x-axis. The perturbation induces an oscillating spin magnetization M_x. If the field has the value $H_x = H_1 \cos \omega t$, the susceptibilities are defined as in the nuclear case by the relation

$$M_x = H_1\{\chi_S''(\omega)\sin \omega t + \chi_S'(\omega)\cos \omega t\}. \tag{3.25}$$

These susceptibilities have a tensorial character and eqn (3.25) is only the definition of the χ_{xx}' and χ_{xx}'' components. In metals the electronic Zeeman energy is usually an isotropic quantity and this energy does not depend on the relative orientation between the field and the crystalline axis. In these conditions it can be shown that all the other components of the susceptibility tensor are functions of χ_S'' and χ_S' defined

by eqn (3.25). (These relations are also valid for the nuclear susceptibilities, see Ref. 1, p. 49.) These relations are

$$\chi''_{xx} = \chi''_{yy}, \qquad \chi'_{xx} = \chi'_{yy},$$

and
$$\chi'_{yx} = -\chi''_{xx}, \qquad \chi''_{yx} = \chi'_{xx}.$$

The two last relations are a consequence of the fact that an isotropic
spin system is only sensitive to the rotating part of the exciting field. As
for the nuclear case, χ'_{S} and χ''_{S} obey the Kramers–Krönig relations.

3.3. Non-local susceptibilities

In Chapter 2, it was shown that the main term of the hyperfine
coupling is the contact term. For studying the effect of this term on
the electronic system it is convenient to introduce the concept of a
susceptibility varying in space. Also in an electron-resonance experiment, because of the occurrence of the skin depth, the electronic system
is submitted to a spatially inhomogeneous excitation. In these two
problems the fields vary in space over a very different scale; the nuclear
interaction varies on an atomic scale, whereas the skin-depth effect
involves a variation over a macroscopic distance.

Let us consider the effect of a magnetic field $\mathbf{H}(\mathbf{R}')$, which varies in
space. A susceptibility $\chi_{S}(\mathbf{R}, \mathbf{R}')$ is defined as follows:

$$\mathbf{M}(\mathbf{R}) = \frac{1}{V} \int \chi_{S}(\mathbf{R}, \mathbf{R}')\mathbf{H}(\mathbf{R}')\, \mathrm{d}^3 R',$$

where $\mathbf{M}(\mathbf{R})$ is the local spin magnetization at the point $\mathbf{R}$ induced by
the perturbing field $\mathbf{H}(\mathbf{R}')$.

The local spin magnetization is defined as the thermal average value
of the operator $g\beta\mathbf{S}(\mathbf{R})$

$$\mathbf{M}(\mathbf{R}) = g\beta\langle\mathbf{S}(\mathbf{R})\rangle \tag{3.26}$$

with
$$\mathbf{S}(\mathbf{R}) = \sum_{e} \mathbf{s}_e\, \delta(\mathbf{r}_e - \mathbf{R}).$$

If one considers the case of a purely local magnetic perturbation

$$\mathbf{H}(\mathbf{R}') = V\mathbf{h}\delta(\mathbf{R}' - \mathbf{R}_i),$$

the local magnetization is simply proportional to $\mathbf{h}$:

$$\mathbf{M}(\mathbf{R}) = \mathbf{h}\chi_{S}(\mathbf{R}, \mathbf{R}_i).$$

If the vector joining the points $\mathbf{R}$ and $\mathbf{R}'$ is a lattice translation the
susceptibility is only a function of the difference vector $\mathbf{R} - \mathbf{R}'$; this
property is valid for all $\mathbf{R}$ and $\mathbf{R}'$ for the electron gas. If we consider only
the susceptibilities having this property, a spatial Fourier transform
$\chi_{S}(\mathbf{q})$ of $\chi_{S}(\mathbf{R} - \mathbf{R}')$ can be defined. This quantity may be interpreted

physically as giving the response of the electronic spin system to a perturbing field of the form

$$\mathbf{H(R')} = \mathbf{h_q}\, e^{i\mathbf{q}.\mathbf{R'}}.$$

Such a perturbation produces an oscillating magnetization

$$\mathbf{M(R)} = \mathbf{M_q}\, e^{i\mathbf{q}.\mathbf{R}}$$

and the following relation may be obtained:

$$\mathbf{M_q} = \chi_S(\mathbf{q})\mathbf{h_q}.$$

Let us calculate $\chi_S(\mathbf{q})$ for a free-electron gas using a perturbation technique.

The problem is to calculate the amplitude $\mathbf{M_q}$ in the presence of the perturbing Hamiltonian $\mathscr{H}_q$.

$$\mathscr{H}_q = -g\beta \int \mathbf{S(R)}.\mathbf{H(R)}\, \mathrm{d}^3 R$$

or

$$\mathscr{H}_q = -g\beta \mathbf{h_q}. \int \mathbf{S(R)} e^{i\mathbf{q}.\mathbf{R}}\, \mathrm{d}^3 R,$$

which will be written more briefly

$$\mathscr{H}_q = -g\beta \mathbf{h_q}.\mathbf{S_{-q}}.$$

As we mentioned, for χ'' (and χ'), $\chi_S(\mathbf{q})$ has a tensorial character and in general depends on the orientation of the vectors $\mathbf{M_q}$ and $\mathbf{h_q}$, but again because of the weakness of the electronic Zeeman energy this complication can be very often forgotten and $\chi_S(\mathbf{q})$ is a diagonal tensor with the three diagonal values equal provided the fields do not vary in time. A simple method for explaining this result is as follows. In all calculations the electronic Zeeman energy will be neglected, thus no change will appear in the value of $\chi_S(\mathbf{q})$ if the static magnetic field is reduced to zero, and in this condition the electron-gas Hamiltonian is invariant by rotation. (This result is still valid if the interactions are taken into account but the situation will be more complicated in the presence of a lattice.) In performing the calculation it is convenient to express the operator $\mathbf{S(R)}$ and $\mathbf{S_q}$ using the second quantization representation. Using eqns (3.7) and (3.8) we find

$$S_z(\mathbf{R}) = \tfrac{1}{2}\sum_{\mathbf{k},\mathbf{k'}} e^{i(\mathbf{k'}-\mathbf{k}).\mathbf{R}}(C^+_{\mathbf{k}\uparrow} C_{\mathbf{k'}\uparrow} - C^+_{\mathbf{k}\downarrow} C_{\mathbf{k'}\downarrow}) \tag{3.27}$$

and

$$S_+(\mathbf{R}) = \sum_{\mathbf{k},\mathbf{k'}} e^{i(\mathbf{k'}-\mathbf{k}).\mathbf{R}} C^+_{\mathbf{k}\uparrow} C_{\mathbf{k'}\downarrow}. \tag{3.28}$$

From these equations the Fourier transforms are easily obtained, for example,

$$S_{z,\mathbf{q}} = \frac{(2\pi)^3}{2}\sum_{\mathbf{k}} (C^+_{\mathbf{k}\uparrow} C_{\mathbf{k+q}\uparrow} - C^+_{\mathbf{k}\downarrow} C_{\mathbf{k+q}\downarrow}). \tag{3.29}$$

Let us start the calculation of $\chi_S(\mathbf{q})$. The perturbing Hamiltonian $\mathscr{H}_{\mathbf{q}}$ admixes the electronic ground state $|G\rangle$ with excited states where an electron $\mathbf{k}-\mathbf{q}$ has been destroyed while an electron $\mathbf{k}$ has been created; the perturbed wave function $|G'\rangle$ may be written as

$$|G'\rangle = |G\rangle - g\beta \frac{(2\pi)^3}{2} h_{z\mathbf{q}} \sum_{\mathbf{k}} \frac{1}{E_{\mathbf{k}\uparrow} - E_{\mathbf{k}-\mathbf{q}\uparrow}} \; C^{+}_{\mathbf{k}\uparrow} C_{\mathbf{k}-\mathbf{q}\uparrow} |G\rangle +$$

$$+ g\beta \frac{(2\pi)^3}{2} h_{z\mathbf{q}} \sum_{\mathbf{k}} \frac{1}{E_{\mathbf{k}\downarrow} - E_{\mathbf{k}-\mathbf{q}\downarrow}} \; C^{+}_{\mathbf{k}\downarrow} C_{\mathbf{k}-\mathbf{q}\downarrow} |G\rangle.$$

We assume that the field $\mathbf{h_q}$ is directed along the z-axis.

The amplitude $\mathbf{M_q}$ is proportional to the average value of the operator $\mathbf{S_q}$, so we have now to take the matrix element of $S_{z,\mathbf{q}}$ using the wave function $|G'\rangle$: a typical element is of the form

$$\simeq h_{z\mathbf{q}} \left\langle G \left| \sum_{\mathbf{k},\mathbf{k}'} C^{+}_{\mathbf{k}'\uparrow} C_{\mathbf{k}'+\mathbf{q}\uparrow} \frac{1}{E_{\mathbf{k}\uparrow} - E_{\mathbf{k}-\mathbf{q}\uparrow}} \; C^{+}_{\mathbf{k}\uparrow} C_{\mathbf{k}-\mathbf{q}\uparrow} \right| G \right\rangle.$$

This calculation may be extended to finite temperature. A typical element of M_{qz} has now the form

$$\simeq h_{z\mathbf{q}} \sum_{i} P_i \left\langle G_i \left| \sum_{\mathbf{k},\mathbf{k}'} C^{+}_{\mathbf{k}'\uparrow} C_{\mathbf{k}'+\mathbf{q}\uparrow} \frac{1}{E_{\mathbf{k}\uparrow} - E_{\mathbf{k}-\mathbf{q}\uparrow}} \; C^{+}_{\mathbf{k}\uparrow} C_{\mathbf{k}-\mathbf{q}\uparrow} \right| G_i \right\rangle,$$

where G_i is any of the N electron states and P_i the statistical weight of this state.

If the average value of the operators is calculated using eqn (3.9), and taking into account the anticommutation relations, we get the result

$$\chi_S(\mathbf{q}) = \tfrac{1}{4}(g\beta)^2 \sum_{\mathbf{k}} \frac{n(E_{\mathbf{k}+\mathbf{q}}) - n(E_{\mathbf{k}})}{E_{\mathbf{k}} - E_{\mathbf{k}+\mathbf{q}}} \tag{3.30}$$

(the Zeeman energies are neglected). If for $E_{\mathbf{k}}$ eqn (3.1) is used the summation over $\mathbf{k}$ can be performed (for $T = 0$ K) leading to the result

$$\chi_S(\mathbf{q}) = \frac{\chi_S}{2}\left\{1 + \frac{(2k_F)^2 - q^2}{4k_F q} \ln \left| \frac{2k_F + q}{2k_F - q} \right| \right\}. \tag{3.31}$$

This function is shown on Fig. 3.3: it is a decreasing function of q starting from the value χ_S at $q = 0$. For $q = 2k_F$, $\chi_S(2k_F) = \chi_S/2$ and the derivative presents an infinite slope. It may be sometimes convenient to define the complex susceptibility $\chi_S(\mathbf{q}, \omega)$ which gives the response of the system to a magnetic perturbation varying in space and in time (see Appendix 1).

The last step will be to calculate $\chi_S(\mathbf{q})$ using the Landau model. This problem is more difficult. In order to perform the calculation the Landau theory must be extended to describe a situation where the quantities $\delta n_{\mathbf{k}\uparrow}$ vary in space. As the Landau model is valid only when $\mathbf{k}$ is in the vicinity of k_{F}, the spatial variation must be slow compared to $1/k_{\mathrm{F}}$, thus the calculation of $\chi_{\mathrm{S}}(q)$ has a meaning only when $q \ll k_{\mathrm{F}}$. This calculation will be presented in the last chapter.

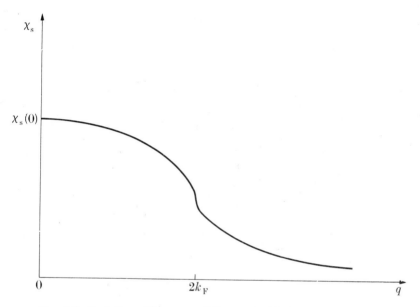

FIG. 3.3. Variation of the susceptibility $\chi_{\mathrm{S}}(q)$ with the wave vector $\mathbf{q}$.

Unfortunately, when we are dealing with nuclear resonance properties such as the relaxation rate we are interested in the susceptibility $\chi_{\mathrm{S}}(\mathbf{R}_i, \mathbf{R}_i)$ and therefore in the value of $\chi_{\mathrm{S}}(\mathbf{q})$ for rather large value of $\mathbf{q}$. For this region there is an estimate of the susceptibility due to P. A. Wolff[16] which is very often used because the result is very simple. He assumes that the range of the electronic interaction is very short (more precisely the electrons are assumed to interact only when they are in contact), consequently the matrix element of this interaction will not depend on the difference between the wave vectors $\mathbf{k}$ and $\mathbf{k}'$ of the electron. With this assumption $\chi_{\mathrm{S}}(q)$ is obtained:

$$\frac{1}{\chi_{\mathrm{S}}(q)} = \frac{1}{\chi_{\mathrm{S}}^0(q)}\left\{1 + B_0 \frac{\chi_{\mathrm{S}}^0(q)}{\chi_{\mathrm{S}}^0(0)}\right\}, \tag{3.32}$$

where B_0 is as defined in eqn (3.23). $B_0 = 2f_{\mathrm{e}}\chi_{\mathrm{S}}^0/(g\beta)^2$ (f_{e} does not now depend upon the vector $\mathbf{k} - \mathbf{k}'$), χ_{S}^0 is the susceptibility in the absence of the interactions which varies as predicted by eqn (3.31). (In $\chi_S^0(q)$ a possible change in the effective mass is taken into account.)

The calculation of $\chi_S(\mathbf{q})$ is of course more difficult if the lattice is taken into account. Equations (3.27), (3.28), and (3.29) are easily generalized. The values of the periodic part of the wave functions $U_\mathbf{k}(\mathbf{R})$ and $U_{\mathbf{k}'}(\mathbf{R})$ appear in the equations. Equation (3.31) for $\chi_S(\mathbf{q})$ is easily generalized but usually the summation over $\mathbf{k}$ is impossible to perform exactly except by a numerical calculation. The susceptibility is a tensor, whose components depend upon the direction of the vector $\mathbf{q}$ in the reciprocal space.

THEORETICAL STUDY OF THE NUCLEAR RESONANCE PROPERTIES OF METALS

In this chapter we shall review the properties of the nuclear spin system that depend on the metallic character. The discussion will include all the information about the metallic state that can be learned from a nuclear resonance experiment.

1. The intensity of the resonance line

The intensity of the resonance line is defined as follows:

$$I_n = \int_0^\infty \chi_n''(\omega) \, d\omega. \tag{4.1}$$

This quantity can be expressed as a function of the static nuclear spin susceptibility using one of the Kramers–Krönig relations:

$$\chi_n'(\omega) - \chi_n'(\infty) = \frac{1}{\pi} \mathscr{P} \int_{-\infty}^{+\infty} \frac{\chi_n''(\omega')}{\omega' - \omega} \, d\omega', \tag{4.2}$$

where the symbol $\mathscr{P}$ means that the principal part of the integral must be taken.

For the special value $\omega = 0$ we get

$$\chi_n'(0) = \chi_n = \frac{1}{\pi} \mathscr{P} \int_{-\infty}^{+\infty} \frac{\chi_n''(\omega')}{\omega'} \, d\omega'. \tag{4.3}$$

(We have assumed $\chi_n'(\infty) = 0$.) The function χ'' is an odd function of ω, and as this function is important only when ω' is in the vicinity of ω_n eqn (4.3) becomes

$$\chi_n = \frac{2}{\pi} \frac{I_n}{\omega_n}. \tag{4.4}$$

Therefore the measurement of the intensity of the line is completely equivalent to a measurement of the static nuclear susceptibility.

This calculation is not limited to a system of nuclear spins and is valid for the electronic system as well. The knowledge of I_n or χ_n is not very interesting because χ_n is known (eqn (1.3)) as a function of N, I, T, and γ_n.

D

In some circumstances, however, the measured value of I_n differs from the value predicted using eqn (4.4). For instance, in the case of a spin I whose value is larger than $I = \frac{1}{2}$, some components of the line are broadened by quadrupolar effects and become unobservable (see Chapters 5 and 7). The measurement of I_n then gives us information about the magnitude of these quadrupolar effects.

There is another effect that tends to reduce the intensity of the line. Because a metal is a conductor the radiofrequency field is affected by the skin effect, which produces an inhomogeneity of the amplitude of the exciting field (and also a change in the phase) if the particles are too large. The calculation of the power absorbed by the nuclear system and of the line shape, taking the skin depth into account, will be presented in Chapter 10.

Let us finally note that in some of the experiments we shall describe using large radiofrequency fields the line shape itself is a function of H_1 and it is very important for interpreting the results to have a uniform exciting field and therefore to use small particles.

2. The position of the line: the Knight shift

2.1. *The contact term*

Conduction electrons produce at the nuclear position an average static magnetic field. This field shifts the nuclear resonance frequency. The displacement is called the 'Knight shift'.

Let us first consider the field due to the contact term. This term produces a field $\mathbf{H}_{ec}$, given by

$$\mathbf{H}_{ec} = -\frac{8\pi}{3} g\beta \mathbf{S}(\mathbf{R}_i),\tag{4.5}$$

where $\mathbf{R}_i$ is the position of the nuclei. The Knight shift is obtained by taking the thermal average of this operator. For metals having a large Fermi energy an extremely good approximation is obtained by taking the average value in the electronic ground state.

Using the second quantization formalism the components of $\mathbf{S}(\mathbf{R}_i)$ may be written (these equations are a generalization of eqns (3.7) and (3.8) established with plane wave functions)

$$S_z(\mathbf{R}_i) = \tfrac{1}{2}\sum_{\mathbf{k},\mathbf{k}'} e^{i(\mathbf{k}'-\mathbf{k})\cdot\mathbf{R}_i} U_{\mathbf{k}}^*(\mathbf{R}_i) U_{\mathbf{k}'}(\mathbf{R}_i)(C_{\mathbf{k}\uparrow}^+ C_{\mathbf{k}'\uparrow} - C_{\mathbf{k}\downarrow}^+ C_{\mathbf{k}'\downarrow}),$$

and a similar equation for $S_+(\mathbf{R}_i)$. The band index n is omitted because, as we shall show, only the electrons at the Fermi surface contribute to the average value.

In the presence of a magnetic field along the z-axis only $S_z(\mathbf{R}_i)$ has a non-zero average value which is given by the equation (using (3.9))

$$\langle S_z(\mathbf{R}_i)\rangle = \tfrac{1}{2}\sum_{\mathbf{k}}|U_{\mathbf{k}}(\mathbf{R}_i)|^2\{n(E_{\mathbf{k}\uparrow})-n(E_{\mathbf{k}\downarrow})\}.$$

As the occupation numbers are different only in the vicinity of the Fermi energy, where states occupied for one of the spin orientations and not for the other are found, the equation can be simplified by considering only the wave function when $\mathbf{k} = \mathbf{k}_{\mathrm{F}}$:

$$\langle S_z(\mathbf{R}_i)\rangle = \tfrac{1}{2}|U_{k_{\mathrm{F}}}(\mathbf{R}_i)|^2\sum_{\mathbf{k}}\{n(E_{\mathbf{k}\uparrow})-n(E_{\mathbf{k}\downarrow})\}. \tag{4.6}$$

Using the notation $U_{k_{\mathrm{F}}}$ we implicitly assume that the wave function $U_{k_{\mathrm{F}}}(\mathbf{R}_i)$ does not depend on the orientation of the wave vector $\mathbf{k}_{\mathrm{F}}$. If this assumption is not valid in eqn (4.6) a better approximation is obtained by taking the average value of $|U_{k_{\mathrm{F}}}(R_i)|^2$ over the Fermi surface. (We shall keep the notation $|U_{k_{\mathrm{F}}}(R_i)|^2$ which will mean the average over the $\mathbf{k}_{\mathrm{F}}$ orientation.) The summation over $\mathbf{k}$ in (4.6) may be expressed using the susceptibility, in which case the average field becomes

$$\langle H_{ez}\rangle = \frac{8\pi}{3}\,\chi_{\mathrm{S}}\,H_0|U_{k_{\mathrm{F}}}(R_i)|^2$$

or $$K = \frac{\langle H_{ez}\rangle}{H_0} = \frac{8\pi}{3}\,\chi_{\mathrm{S}}'|U_{k_{\mathrm{F}}}(R_i)|^2 V = \frac{8\pi}{3}\,\chi_{\mathrm{S}}'\,\Omega|\psi_{k_{\mathrm{F}}}(R_i)|^2 \tag{4.7}$$

(see eqns (3.13) and (3.14)), where χ_{S}' is the susceptibility per unit volume. It is quite important to notice that the only assumption made for deriving eqn (4.7) was that unpaired electrons are found only if their energy is in the vicinity of the Fermi energy. The equation is valid in the presence of electron–electron interactions and even in the case of a superconducting metal.

2.2. *The shift of the electronic resonance*

Conversely, the nuclear moments induce a shift of the electronic resonance line. The properties of electronic resonance will be studied in Chapter 10, but it is possible to calculate now the shift of this line due to the nuclear field. Electronic resonance involves the motion of the total spin magnetization and therefore depends only on the behaviour of electrons in the vicinity of the Fermi level.

For these electrons the average values of the operators S_z and $S_z(\mathbf{R}_i)$ are simply proportional (this result is also valid for the other components (4.6))

$$|U_{k_{\mathrm{F}}}(\mathbf{R}_i)|^2\langle S_z\rangle = \langle S_z(\mathbf{R}_i)\rangle$$

and the averaged contact hyperfine coupling may be written

$$\mathscr{H}_n = \hbar\gamma_n \, g\beta\mathbf{I}_i \cdot \langle\mathbf{S}\rangle |U_{k_F}(\mathbf{R}_i)|^2 \, \frac{8\pi}{3}.$$

Such a term produces a shift of the resonance line having the value

$$\langle H_{nz}\rangle = -\hbar\gamma_n \frac{8\pi}{3}\langle I_z\rangle |U_{k_F}(\mathbf{R}_i)|^2, \tag{4.8}$$

where $\langle I_z\rangle$ is the total average magnetization that can be written as a function of the nuclear susceptibility (1.3). As this quantity is always very small, the shift is observable only if the electronic resonance line is very narrow. Knowledge of this field would be very interesting because the nuclear susceptibility is a well known quantity and $\langle H_{nz}\rangle$ provides a direct measurement of $|U_{k_F}(\mathbf{R}_i)|^2$, that is to say of the density at the nucleus for electrons having the Fermi energy. The interpretation of the observed Knight shift is not so easy in general because the electronic susceptibility is known for three metals only.

2.3. *The dipolar term*

The dipolar term in the hyperfine interaction will be considered now. This term produces the average field $\langle\mathbf{H}_{ed}\rangle$ given by the relation

$$\langle\mathbf{H}_{ed}\rangle = -g\beta\Big\langle \sum_e \Big(\frac{-\mathbf{s}_e}{|\mathbf{r}_e-\mathbf{R}_i|^3}+3\,\frac{(\mathbf{r}_e-\mathbf{R}_e)\cdot[\mathbf{s}_e\cdot(\mathbf{r}_e-\mathbf{R}_i)]}{|\mathbf{r}_e-\mathbf{R}_i|^5}\Big)\Big\rangle. \tag{4.9}$$

As in the previous calculation, only the z-component of $\mathbf{s}_e$ has a non-vanishing average value. Using the second quantization this field may be written

$$\langle H^\alpha_{ed}\rangle = -g\beta\Big\langle \sum_{\beta,\mathbf{k},\mathbf{k'}} \mathscr{T}^{\alpha\beta}_{\mathbf{kk'}} \, s^\beta_{\mathbf{kk'}}\Big\rangle \quad (\alpha,\beta = x,y,\text{ or } z), \tag{4.10}$$

where
$$s^z_{\mathbf{kk'}} = \tfrac{1}{2}(C^+_{\mathbf{k}\uparrow} C_{\mathbf{k'}\uparrow} - C^+_{\mathbf{k}\downarrow} C_{\mathbf{k'}\downarrow}),$$

$$s^x_{\mathbf{kk'}}+is^y_{\mathbf{kk'}} = C^+_{\mathbf{k}\uparrow} C_{\mathbf{k'}\downarrow},$$

and
$$\mathscr{T}^{\alpha\beta}_{\mathbf{kk'}} = \int U^*_{\mathbf{k}}(\mathbf{R})U_{\mathbf{k'}}(\mathbf{R})e^{i(\mathbf{k'}-\mathbf{k})\cdot\mathbf{R}} W^{\alpha\beta}(\mathbf{R}) \, d^3R, \tag{4.11}$$

with
$$W^{\alpha\beta}(\mathbf{R}) = \frac{\delta_{\alpha\beta}}{|\mathbf{R}-\mathbf{R}_i|^3} - 3\,\frac{(A-A_i)(B-B_i)}{|\mathbf{R}-\mathbf{R}_i|^5}, \tag{4.12}$$

A_i, B_i, C_i being equal to the projections of $\mathbf{R}_i$ on to the coordinate axes α, β, γ and A, B, C being the projections of $\mathbf{R}$.

Knowing the quantities $\mathscr{T}^{\alpha\beta}_{\mathbf{kk'}}$ the average value of H_{ed} is obtained:

$$\langle H^\alpha_{ed}\rangle = \mathscr{T}^{\alpha z}_{k_F k_F} \chi_S H_0. \tag{4.13}$$

As an example,

$$\mathcal{T}_{k_\mathrm{F}k_\mathrm{F}}^{zz} = \int |U_{k_\mathrm{F}}(\mathbf{R})|^2 \left(\frac{1-3\cos^2\theta_i}{|\mathbf{R}-\mathbf{R}_i|^3}\right) \mathrm{d}^3 R,$$

where θ_i is the angle between $\mathbf{R}-\mathbf{R}_i$ and the z-axis. (As for the contact term the notation k_F means that an average over $\mathbf{k}_\mathrm{F}$ is taken.)

To simplify the notation, $\mathcal{T}_{k_\mathrm{F}k_\mathrm{F}}^{\alpha\beta}$ will be written as $\mathcal{T}_\mathrm{F}^{\alpha\beta}$.

Instead of taking the z-axis along the direction of the field, it may be more convenient to take a system of axes where the tensor $\mathcal{T}_\mathrm{F}^{\alpha\beta}$ has a simpler form. As $\mathcal{T}_\mathrm{F}^{\alpha\beta}$ is a symmetric tensor, it can be diagonalized by a suitable choice of the coordinate axes, which are called X, Y, Z. Only three quantities appear in the calculation: $\mathcal{T}_\mathrm{F}^{XX}$, $\mathcal{T}_\mathrm{F}^{YY}$, and $\mathcal{T}_\mathrm{F}^{ZZ}$. The interaction between the nuclear moment and the average field becomes

$$\mathcal{H}_\mathrm{int} = -\hbar\gamma_\mathrm{n} g\beta(\mathcal{T}_\mathrm{F}^{XX}I_X\langle S_X\rangle + \mathcal{T}_\mathrm{F}^{YY}I_Y\langle S_Y\rangle + \mathcal{T}_\mathrm{F}^{ZZ}I_Z\langle S_Z\rangle). \quad (4.14)$$

If the symmetry at the nuclear site is cubic, the three terms of the tensor are equal and since the dipolar interaction is traceless

$$(W^{XX} + W^{YY} + W^{ZZ} = 0)$$

the interaction disappears.

Let us call θ and ϕ the polar angles of the field direction with respect to the new system of coordinates. As the dipolar field seen by the nuclear moment is much smaller than the external field, only its component along the external field will contribute to the shift and this averaged component is given by the equation

$$H_\mathrm{ed}(\text{along } \mathbf{H}_0) = g\beta(\mathcal{T}_\mathrm{F}^{ZZ}\cos^2\theta + \mathcal{T}_\mathrm{F}^{XX}\sin^2\theta\cos^2\phi +$$
$$+ \mathcal{T}_\mathrm{F}^{YY}\sin^2\theta\sin^2\phi)\langle S_z\rangle, \quad (4.15)$$

or, taking into account the traceless character of the tensor,

$$H_\mathrm{ed}(\text{along } \mathbf{H}_0) = g\beta\mathcal{T}_\mathrm{F}^{ZZ}\langle S_z\rangle\left\{\tfrac{1}{2}(3\cos^2\theta - 1) + \frac{\eta_\mathrm{d}}{2}\sin^2\theta\cos 2\phi\right\}$$

$$(4.16)$$

with

$$\eta_\mathrm{d} = (\mathcal{T}_\mathrm{F}^{XX} - \mathcal{T}_\mathrm{F}^{YY})/\mathcal{T}_\mathrm{F}^{ZZ}.$$

If the wave functions at the Fermi surface are known, we can calculate the two quantities $\mathcal{T}_\mathrm{F}^{ZZ}$ and η_d. As the components of the tensor are averages of second-order spherical harmonics, it is convenient to expand the wave function $U_{k_\mathrm{F}}(\mathbf{R})$ in spherical harmonics. The isotropic s-parts of this expansion will not contribute to the dipolar field.

Going back to eqns (4.11) and (4.12) where the tensor $\mathcal{T}_\mathrm{F}^{\alpha\beta}$ is defined, it may be convenient for the integration to consider two parts. The first comes from the integration over the unit cell centred around the nucleus

i, and is large because $|\mathbf{R}_i - \mathbf{R}|$ is small. The second part will be due to the integration over the volume of the unit cells around all other nuclei j. The total integral may be written

$$\mathscr{T}_{\mathrm{F}}^{ZZ} = \sum_j \mathscr{T}_{ij}^{ZZ},$$

with the definition

$$\mathscr{T}_{ij}^{ZZ} = \int\limits_{\text{cell } j} |U_{k_{\mathrm{F}}}(\mathbf{R})|^2 \frac{(1 - 3 \cos^2\theta_i)}{|\mathbf{R} - \mathbf{R}_i|^3} \, \mathrm{d}^3 R.$$

Let us calculate $\mathscr{T}_{ij}^{ZZ}$ for $i \neq j$. The function $U_{k_{\mathrm{F}}}(\mathbf{R})$ is large usually only when $\mathbf{R}$ is in the vicinity of $\mathbf{R}_j$, and as $1/|\mathbf{R} - \mathbf{R}_i|^3$ varies slowly over the volume of the cell j (an approximation that may not be very good for the cells in the vicinity of the i cell) this function may be written as

$$\mathscr{T}_{ij}^{ZZ} = \int\limits_{\text{cell } j} |U_{k_{\mathrm{F}}}(\mathbf{R})|^2 \, \mathrm{d}^3 R \, \frac{1 - 3 \cos^2\theta_{ij}}{|\mathbf{R}_j - \mathbf{R}_i|^3}.$$

θ_{ij} is the angle between the Z-axis and the vector $\mathbf{R}_j - \mathbf{R}_i$, and as

$$\int\limits_{\text{cell}} |U_{k_{\mathrm{F}}}(\mathbf{R})|^2 \, \mathrm{d}^3 R = \frac{\Omega}{V},$$

the total contribution $\mathscr{T}_{\mathrm{F}}^{zz}\langle S_z \rangle$ becomes

$$\mathscr{T}_{\mathrm{F}}^{zz}\langle S_z \rangle = \mathscr{T}_{ii}^{zz}\langle S_z \rangle + \frac{\Omega}{V}\langle S_z \rangle \sum_{j \neq i} \frac{1 - 3 \cos^2\theta_{ij}}{|\mathbf{R}_j - \mathbf{R}_i|^2}.$$

The last term has a very simple physical interpretation: it represents the dipolar field due to the electronic magnetization assumed to be localized at the centre of each cell. As is well known, even if the symmetry at the nucleus is cubic, this term exists if the sample is not spherical. This field is the demagnetizing field for a system having the magnetization M_{S}.

This calculation is somewhat academic because, as stated in the previous chapter, the magnetization of the electronic system has other components of diamagnetic origin of a comparable magnitude which also produces demagnetizing fields. For metals with large susceptibility (as the transition metals) the shifts due to the demagnetizing field are not negligible.[17] As the experiments are done using powder of non-definite shape an inhomogeneous broadening is observed.

2.4. *The orbital shift*

It has been already noted that the average value of the orbital moment is usually equal to zero. However, as in metals the electrons are not localized, the currents produced by the presence of a static magnetic field are important and this problem requires a more detailed discussion.

It is well known that even in an insulating non-magnetic substance there exist shifts having an orbital origin (see Ref. 1, p. 175). These shifts, called chemical shifts, are also present in metals because of the existence of the inner shell electrons. However, the chemical shifts are usually much smaller than the Knight shifts. The calculations of the orbital field using localized atomic orbitals are not very simple and of course the evaluation in a metal is even more difficult (and consequently rather uncertain).

In Chapter 2 the hyperfine orbital coupling was written as

$$\mathscr{H}_{\mathrm{ho}} = \hbar\gamma_n g\beta \mathbf{I}_i \cdot \sum_e \frac{\mathbf{l}_{ie}}{|\mathbf{r}_e - \mathbf{R}_i|^3},$$

with
$$\hbar\mathbf{l}_{ie} = (\mathbf{r}_e - \mathbf{R}_i) \wedge \mathbf{p}_e.$$

In the presence of a magnetic field the quantity $\mathbf{p}$ in the Hamiltonian is replaced by $\mathbf{p} + (e/c)\mathbf{A}$, where $\mathbf{A}$ is the vector potential that obeys the two equations
$$\mathbf{H} = \mathrm{curl}\ \mathbf{A}, \qquad \mathrm{div}\ \mathbf{A} = 0.$$

If in the interaction $\mathscr{H}_{\mathrm{ho}}$, $\mathbf{p}$ is replaced by its new value, a contribution proportional to the external field appears. This contribution is interpreted physically as describing the interaction between the nuclear moment and the currents induced by the electronic precession around the field $\mathbf{H}_0$.

But we have the usual difficulty due to the fact that $\mathbf{A}$ is not completely defined. For electrons localized around a nucleus i a reasonable choice is
$$\mathbf{A} = \tfrac{1}{2}\mathbf{H}_0 \wedge (\mathbf{r} - \mathbf{R}_i).$$

In a metal the choice of a gauge is not obvious and for the moment we shall write
$$\mathbf{A} = \tfrac{1}{2}\mathbf{H}_0 \wedge (\mathbf{r} - \mathbf{R}).$$

$\mathbf{R}$ is an unspecified vector we shall choose later.

Taking the average value of the term linear in H_0 in $\mathscr{H}_{\mathrm{ho}}$ over the electronic ground state wave function, the following orbital shift is obtained:

$$H^1_{0z} = -\frac{g\beta e}{2c} H_0 \sum_{\mathbf{k},n} n(E_{\mathbf{k},n}) L^1_{\mathbf{k}n,\mathbf{k}n,z} \qquad (4.17)$$

with
$$L^1_{\mathbf{k}n,\mathbf{k}n,z} = \int U^*_{\mathbf{k}n}(\mathbf{r}) \frac{(x-X)(x-X_i)+(y-Y)(y-Y_i)}{|\mathbf{r}-\mathbf{R}_i|^3} U_{\mathbf{k}n}(\mathbf{r})\ \mathrm{d}^3r.$$

This result so obtained is not very satisfactory because H^1_{0z} may be arbitrarily changed by changing $\mathbf{R}$, a result which violates the law that an observable must be gauge invariant.

There is another orbital contribution to the shift which is due to the following effect. In the total Hamiltonian of the system there are terms omitted so far which describe the coupling between $\mathbf{p}_e$ and the vector potential. Let us consider for simplicity the Hamiltonian for one electron. The kinetic energy term is $(1/2m)\{\mathbf{p}+(e/c)\mathbf{A}\}^2$ and the term linear in $\mathbf{A}$ may be written

$$\mathscr{H}_0 = \frac{e}{2mc}\,(\mathbf{p}.\mathbf{A}+\mathbf{A}.\mathbf{p}) = \beta\mathbf{L}.\mathbf{H}_0, \qquad (4.18)$$

with $\mathbf{L}$ defined as

$$\hbar\mathbf{L} = \tfrac{1}{2}\{\mathbf{p}\,\wedge\,(\mathbf{r}-\mathbf{R})+(\mathbf{r}-\mathbf{R})\wedge\mathbf{p}\}.$$

This interaction slightly admixes the wave functions of the various bands and the average value of the hyperfine orbital coupling operator $\mathscr{H}_{\mathrm{ho}}$ (neglecting now $\{e/c\}\mathbf{A}$ which leads to higher-order corrections) is no longer equal to zero. The Hamiltonian (4.18) for the N-electron system may be written

$$\mathscr{H}_0 = \beta\mathbf{H}_0 . \sum_{\mathbf{k},\mathbf{k}',n,n'} C^+_{\mathbf{k}n} C_{\mathbf{k}'n'}\,{}^2\mathbf{L}_{\mathbf{k}'n',\mathbf{k}n},$$

where

$${}^2\mathbf{L}_{\mathbf{k}'n',\mathbf{k}n} = \int U^*_{\mathbf{k}n}(\mathbf{r})\,e^{i\mathbf{k}.\mathbf{r}}\mathbf{L}\,e^{i\mathbf{k}'.\mathbf{r}}U_{\mathbf{k}'n'}(\mathbf{r})\,d^3r \qquad (4.18')$$

and for the Hamiltonian $\mathscr{H}_{\mathrm{ho}}$:

$$\mathscr{H}_{\mathrm{ho}} = \hbar\gamma_{\mathrm{n}}\,\mathbf{H}_{\mathrm{co}}.\mathbf{I}, \qquad (4.19)$$

where

$$\mathbf{H}_{\mathrm{co}} = -g\beta\sum_{\mathbf{k}n,\mathbf{k}'n'}\mathscr{L}^i_{\mathbf{k}'n',\mathbf{k}n}\,C^+_{\mathbf{k}n}\,C_{\mathbf{k}'n'}$$

with

$$\mathscr{L}^i_{\mathbf{k}'n',\mathbf{k}n} = \int U^*_{\mathbf{k}n}\,e^{-i\mathbf{k}.\mathbf{r}}\,\frac{\mathbf{l}_i}{|\mathbf{r}-\mathbf{R}_i|^3}\,e^{i\mathbf{k}'.\mathbf{r}}U_{\mathbf{k}'n'}\,d^3r; \qquad (4.19')$$

using a second-order perturbation calculation we obtain the shift

$$H^2_{0z} = -g\beta^2 H_0\sum_{\mathbf{k}\mathbf{k}',nn'} n(E_{\mathbf{k}'n'})\big(1-n(E_{\mathbf{k}n})\big)\times$$

$$\times\frac{\mathscr{L}^i_{\mathbf{k}'n',\mathbf{k}nz}\,{}^2L_{\mathbf{k}n,\mathbf{k}'n'z}+\mathscr{L}^i_{\mathbf{k}n,\mathbf{k}'n'z}\,{}^2L_{\mathbf{k}'n',\mathbf{k}nz}}{E_{\mathbf{k}',n'}-E_{\mathbf{k},n}}. \qquad (4.20)$$

This complicated equation may be simplified by using the gauge

$$\mathbf{r}-\mathbf{R} = \mathbf{r}-\mathbf{R}_j \qquad (4.21)$$

if $\mathbf{r}$ is in the unit cell around the nucleus j. With such a choice $\mathbf{A}$ and $\mathbf{L}$ are periodic and $\mathbf{L}$ has only matrix elements between wave functions having the same wave vector $\mathbf{k}$.

As in the calculation of the dipolar shift we may consider separately in the calculation of $\mathscr{L}$ and H^1_{0z} the contribution of the integration over the cell around the nucleus and around all the other cells. The contribution of the cell i should be comparable in magnitude to a chemical shift, the functions $U_{\mathbf{k}n}$ having great similarity with atomic wave functions. The evaluation of the other contribution is more difficult. These difficulties are related to the fact that for some metals the average orbital moment becomes large. Such a situation is met in semiconductors where two bands separated by a small energy gap are found. For simple metals with an electronic g-factor for conduction electrons nearly equal to 2 (which means the average orbital moment is not exceedingly large) all these contributions are presumably small. For transition metals the orbital shift becomes larger because two energy bands that do not differ very much in energy are present and the field given by eqn (4.20) becomes important.

3. The relaxation time

3.1. *A simple calculation of the relaxation rate*

In a metal at a low temperature the nuclear relaxation is produced by transitions among the electronic and nuclear energy levels induced by the hyperfine coupling. Let us first consider the influence of the contact term and neglect the interactions among electrons.

The transitions among the electronic energy levels are induced by the operators $S_+(\mathbf{R}_i)$ or $S_-(\mathbf{R}_i)$. First the case of a spin $I = \frac{1}{2}$ will be considered. Equation (1.12) shows that the relaxation time T_1 is defined by the relation $1/T_1 = W_\uparrow + W_\downarrow$. For the moment the difference between the two probabilities will be neglected. The quantity $W_\uparrow$ is due to the transitions induced by the operator $I_+ S_-(\mathbf{R}_i)$ of the coupling. Equation (3.28) shows us that the operator $S_-(\mathbf{R}_i)$ is able to destroy an electron in a state $\mathbf{k}$ with spin up and to create an electron in a state $\mathbf{k}'$ with spin down. More precisely this operator will be written

$$S_-(\mathbf{R}_i) = \sum_{\mathbf{k},\mathbf{k}'} e^{i(\mathbf{k}-\mathbf{k}')\cdot\mathbf{R}_i} U_{\mathbf{k}}(\mathbf{R}_i) U_{\mathbf{k}'}^*(\mathbf{R}_i) C_{\mathbf{k}'\downarrow}^+ C_{\mathbf{k}\uparrow}. \tag{4.22}$$

(Again the band index n is omitted because only the electrons in the vicinity of the Fermi level contribute to the relaxation rate.)

Using eqn (4.22) the probability for destroying the electron $\mathbf{k}\uparrow$, while creating the electron $\mathbf{k}'\downarrow$ and reversing the nuclear spin, is given by the well-known formula

$$W_{\mathbf{k}\uparrow,\mathbf{k}'\downarrow} = \frac{2\pi}{\hbar}\left(\frac{8\pi}{3}\gamma_{\mathrm{n}}g\beta\hbar\right)^2 \tfrac{1}{4}|U_{\mathbf{k}}(\mathbf{R}_i)|^2|U_{\mathbf{k}'}(\mathbf{R}_i)|^2\delta(E_{\mathbf{k}\uparrow}-E_{\mathbf{k}'\downarrow}-\hbar\omega_{\mathrm{n}}). \tag{4.23}$$

This probability has to be multiplied by the probability $n(E_{\mathbf{k}\uparrow})$ of finding the state $\mathbf{k}\uparrow$ occupied and by the probability $\{1-n(E_{\mathbf{k}'\downarrow})\}$ of finding the state $\mathbf{k}'\downarrow$ empty.

The probability $W_\uparrow$ is now obtained by summing over all the possible states $\mathbf{k}$ and $\mathbf{k'}$.

Neglecting the electronic and nuclear Zeeman energy, the relaxation time is given by the equation

$$\frac{1}{T_1} = \frac{64\pi^3}{9}\,(g\beta\gamma_n)^2\hbar \sum_{\mathbf{k},\mathbf{k'}} n(E_\mathbf{k})\{1-n(E_\mathbf{k'})\}\delta(E_\mathbf{k}-E_\mathbf{k'})\times$$
$$\times\{U_\mathbf{k}(\mathbf{R}_i)\}^2\{U_\mathbf{k'}(\mathbf{R}_i)\}^2. \quad (4.24)$$

The summation over the wave vectors $\mathbf{k}$ and $\mathbf{k'}$ may be replaced by an integration over the energies. If it is also assumed that the wave function does not vary appreciably over the Fermi surface, we obtain

$$\frac{1}{T_1} = \frac{64\pi^3}{9}\,(g\beta\gamma_n)^2\hbar \int_0^\infty \int_0^\infty |U_{k_\mathrm{F}}(\mathbf{R}_i)|^4 n(E)\{1-n(E')\}\times$$
$$\times\delta(E-E')\frac{V^2}{4}\,g(E)g(E')\,\mathrm{d}E\mathrm{d}E'. \quad (4.25)$$

(The density of state for a given spin orientation is equal to $\tfrac{1}{2}Vg(E)$.)

Using eqn (3.5) we notice that the product $n(E)\{1-n(E')\}$ is equal to $-k_\mathrm{B}T\,\mathrm{d}n(E)/\mathrm{d}E$. As the temperature is small compared to T_F the function $\mathrm{d}n(E)/\mathrm{d}E$ may be replaced by $-\delta(E-E_\mathrm{F})$.

The following final result is obtained:

$$\frac{1}{T_1} = \frac{16\pi^3}{9}\,(g\beta\gamma_n)^2\hbar|U_{k_\mathrm{F}}(\mathbf{R}_i)|^4 k_\mathrm{B}TV^2\{g(E_\mathrm{F})\}^2. \quad (4.26)$$

This equation is not restricted to the spin value $I = \tfrac{1}{2}$; it can be shown that eqns (1.12) and (4.26) are valid as well for other values of I (see Ref. 3, pp. 125 and 126).

3.2. The Korringa relation

If the Knight shift is calculated using eqn (4.7) and if for χ'_s the value for a free-electron gas is taken (3.19), a very simple relation is found between the relaxation rate and the Knight shift. This relation is called the Korringa relation and is as follows:

$$K^2 = \frac{\hbar}{4\pi k_\mathrm{B}TT_1}\left(\frac{\gamma_\mathrm{e}}{\gamma_\mathrm{n}}\right)^2. \quad (4.27)$$

The Korringa relation is very useful for calculating an estimate of the relaxation rate knowing the value of the Knight shift. Obviously this relation is only an approximation.

The interaction between electrons produces a change in χ_S and also in $g(E_\mathrm{F})$, and a more satisfactory but also more complicated relation is obtained by writing

$$TT_1 K^2 = \frac{\hbar}{4\pi k_\mathrm{B}} \left(\frac{\gamma_\mathrm{e}}{\gamma_\mathrm{n}}\right)^2 \left(\frac{\chi_\mathrm{S}}{\chi_\mathrm{S}^\mathrm{F}}\right)^2 \left\{\frac{g(E_\mathrm{F})}{g^\mathrm{F}(E_\mathrm{F})}\right\}^2, \tag{4.28}$$

where χ_S and $g(E_\mathrm{F})$ are respectively the experimental values of the susceptibility and density of states, whereas $\chi_\mathrm{S}^\mathrm{F}$ and $g^\mathrm{F}(E_\mathrm{F})$ are the calculated values for non-interacting electrons. For a simple metal eqn (4.28) is usually in better agreement with the experimental results than eqn (4.27). However there are other corrections to eqn (4.28). First, we have neglected so far the influence of the interactions on the relaxation rate. Secondly, both in the Knight shift and in the relaxation rate there are contributions coming from the other terms of the hyperfine interaction. The Korringa relation may sometimes be used for evaluating the order of magnitude of these other contributions.

3.3. *Orbital and dipolar relaxation rates*

The calculation of the relaxation rates coming from the other terms of the hyperfine interaction will be performed exactly as for the contact term. (We still neglect the electron–electron interactions.) Let us first consider the orbital coupling. Using eqn (4.19) the relaxation rate is found to be

$$\frac{1}{T_1} = \frac{4\pi}{\hbar}(g\beta\gamma_\mathrm{n}\hbar)^2 \sum_{\mathbf{k},\mathbf{k}'} \tfrac{1}{4}(\mathscr{L}_{\mathbf{k}',\mathbf{k},x} + \mathrm{i}\mathscr{L}_{\mathbf{k}',\mathbf{k},y})^2 \times$$
$$\times \delta(E_\mathbf{k} - E_{\mathbf{k}'} - \hbar\omega_\mathrm{n})n(E_\mathbf{k})\{1 - n(E_{\mathbf{k}'})\}; \tag{4.29}$$

the quantities $\mathscr{L}_{\mathbf{k}',\mathbf{k},x}$ and $\mathscr{L}_{\mathbf{k}',\mathbf{k},y}$ are defined in eqn (4.19′). For the dipolar relaxation rate a similar equation is found, but instead of the $\mathscr{L}_{\mathbf{k}',\mathbf{k}}$ matrix elements we use elements of the tensors $\mathscr{T}_{\mathbf{k},\mathbf{k}'}^{\alpha\beta}$ previously defined.

The two rates are proportional to $k_\mathrm{B} T$ and to the square of the density of states. For calculating the matrix elements, it is necessary to know the value of the wave function at the Fermi level. It is convenient to expand the wave function in spherical harmonics:

$$\Phi_\mathbf{k}(\mathbf{r}) = \Phi_\mathbf{k}^s(r) + \sum_m \Phi_\mathbf{k}^p(r)Y_1^m(\theta,\phi) + \dots .$$

(Note that we expand the function $\Phi_\mathbf{k}$ and not the function $U_\mathbf{k}$ even if $U_\mathbf{k}$ is approximately an s-wave function. $\Phi_\mathbf{k}$ has no reason to have the same property, because the factor $\mathrm{e}^{\mathrm{i}\mathbf{k}.\mathbf{r}}$ must also be expanded.)

The orbital and dipolar contributions are of the same order of magnitude because the matrix elements involved have a comparable form. The s-part of the wave function does not contribute to the rate. If the expansion is limited to the p-wave and if the Fermi surface is assumed to be a sphere, the calculation can be completed.[18] The wave function takes the simple form

$$\Phi_{\mathbf{k}}(r) = f(r) + h(r)\mathbf{k} \cdot \mathbf{r},$$

where f and h are functions of the magnitude of $\mathbf{r}$ only. The matrix elements of the various operators of the hyperfine coupling are then easily calculated. For example, one finds

$$\mathscr{L}_{\mathbf{k'},\mathbf{k},x} = \tfrac{2}{3}(k_y\, k'_z - k'_y\, k_z) \int \frac{|h(r)|^2}{r}\, d^3r.$$

A detailed calculation of these rates was also done by Obata[19] for transition elements using tight binding wave functions. This calculation will be discussed in Chapter 8 and also in Appendix 2. Note that even for a site with cubic symmetry the orbital and dipolar rates exist, in contrast to what happens for the Knight shift due to the dipolar interaction. If the symmetry is lower than cubic these rates depend upon the orientation of the magnetic field with respect to the crystalline axis.

3.4. *The Overhauser effect*[20]

In 1953 Overhauser predicted a very interesting effect coming from the properties of the nuclear relaxation in a metal. It was proved later[21] that this effect is also present in other systems. Overhauser showed that by saturating the electron spin resonance line, the nuclear polarization is changed and, more precisely, considerably enhanced. The Overhauser effect is very interesting because the change in the nuclear polarization is a function of the nature of the relaxation rate. The result of the calculation (which will be explained later) is that the ratio of the populations of two adjacent nuclear sublevels, instead of being given by eqn (1.10), becomes

$$\frac{N_{m-1}}{N_m} = \exp\left\{ -\frac{\hbar \omega_n}{k_\mathrm{B} T} \left(1 + s\xi\, \frac{\gamma_e}{\gamma_n} \right) \right\}. \tag{4.30}$$

s is the degree of saturation of the electronic resonance (s varies between 0 and 1; $s = 1$ means complete saturation) and ξ is a numerical factor which is a function of the various relaxation rates. It can be shown that ξ has the value

$$\xi = \left\{ -\left(\frac{1}{T_1}\right)_\mathrm{contact} + \frac{1}{2}\left(\frac{1}{T_1}\right)_\mathrm{dipolar} \right\} \bigg/ \left(\frac{1}{T_1}\right)_\mathrm{total}, \tag{4.31}$$

where $(1/T_1)_\mathrm{total}$ is the total nuclear relaxation rate including the orbital relaxation and eventually contributions not coming from the magnetic hyperfine interaction. If the temperature is high, that is to say if $k_B T \gg \hbar\omega_0$, eqn (4.30) may be expanded and the nuclear polarization is given by the equation

$$\langle I_z \rangle = \langle I_z \rangle_0 \left(1 + s\xi \frac{\gamma_e}{\gamma_n}\right), \tag{4.32}$$

where $\langle I_z \rangle_0$ is the equilibrium nuclear polarization.

This result will be proved now for the special case of a purely scalar rate ($\xi = -1$) and for non-interacting electrons. (The extension of the calculation for taking the other relaxation mechanism into account is straightforward.)

For an assembly of non-interacting electrons, it is possible to consider that in the presence of the external field, the electrons having different spin orientations are different systems with their own Fermi energies E_F^+ and E_F^-. Of course for an equilibrium situation we have the relation $E_F^+ = E_F^- = E_F$. If the electronic line is completely saturated the magnetization is reduced to zero and the same number of electrons with the two-spin orientations must be realized:

$$\sum_k n_{k\uparrow} = \sum_k n_{k\downarrow}.$$

As the electronic energies are given by the two relations

$$E_{k\uparrow} = E_k + \frac{\hbar\omega_0}{2} \quad \text{and} \quad E_{k\downarrow} = E_k - \frac{\hbar\omega_0}{2}$$

the equality of the total occupation numbers implies the existence of a difference $\hbar\omega_0$ between the two Fermi energies (see Fig. 4.1).

For situations of partial saturation the parameter s will be defined by the relation

$$E_F^+ - E_F^- = s\hbar\omega_0. \tag{4.33}$$

This definition of s is equivalent to the usual definition of the parameter of saturation which is

$$\langle S_z \rangle = \langle S_z \rangle_0 (1 - s), \tag{4.34}$$

where $\langle S_z \rangle_0$ is the equilibrium value for $\langle S_z \rangle$. Let us first show the equivalence of the two definitions of s. The average value of $\langle S_z \rangle$ is given by the integral

$$\langle S_z \rangle \simeq \int_0^\infty g(E_k)[n_\uparrow(E_k) - n_\downarrow(E_k)]\, dE_k,$$

where $n_\uparrow(E_k)$ is the occupation probability for an electron of spin up and with the kinetic energy E_k. The value of this quantity is

$$n_\uparrow(E_k) = \left[\exp\left\{\left(E_k + \frac{\hbar\omega_0}{2} - E_F^+\right) \middle/ k_B T\right\} + 1\right]^{-1}.$$

Similarly,

$$n_\downarrow(E_{\mathbf{k}}) = \left[\exp\left\{\left(E_{\mathbf{k}} - \frac{\hbar\omega_0}{2} - E_{\mathrm{F}}^-\right)\Big/ k_{\mathrm{B}}T\right\} + 1\right]^{-1}.$$

If we introduce the function $\bar{n}(x) = (e^{x/k_{\mathrm{B}}T} + 1)^{-1}$, the value of $\langle S_z \rangle$ may be written

$$\langle S_z \rangle \simeq \int_0^\infty g(E_{\mathbf{k}}) \left\{ \bar{n}\left(E_{\mathbf{k}} - E_{\mathrm{F}}^+ + \frac{\hbar\omega_0}{2}\right) - \bar{n}\left(E_{\mathbf{k}} - E_{\mathrm{F}}^- - \frac{\hbar\omega_0}{2}\right) \right\} dE_{\mathbf{k}};$$

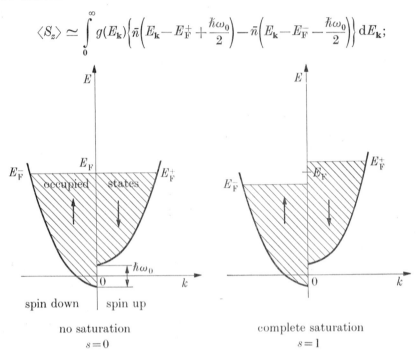

spin down | spin up

no saturation
$s = 0$

complete saturation
$s = 1$

FIG. 4.1. Position of the Fermi energies in the presence of an electronic saturation.

using the relation (4.33), we find

$$\langle S_z \rangle \simeq \int_0^\infty g(E_{\mathbf{k}}) \left[\bar{n}\left\{ E_{\mathbf{k}} - E_{\mathrm{F}} + \frac{\hbar\omega_0}{2}(1-s) \right\} - \bar{n}\left\{ E_{\mathbf{k}} - E_{\mathrm{F}} - \frac{\hbar\omega_0}{2}(1-s) \right\} \right] dE_{\mathbf{k}},$$

with $2E_{\mathrm{F}} = E_{\mathrm{F}}^+ + E_{\mathrm{F}}^-$.

This last equation may be interpreted very simply. The presence of a saturation of the line $s \neq 0$ has simply the effect of changing in the calculation of the magnetization, the Zeeman energy $\hbar\omega_0$, which has to be replaced by $\hbar\omega_0(1-s)$. The same magnetization would be obtained if in an equilibrium situation the external field H_0 becomes $H_0(1-s)$.

This result is true only if the quantity E_{F} we introduce does not vary with the applied field or with the parameter s. By writing that the total number of electrons N does not depend on s or H_0 $\left(\dfrac{\partial N}{\partial s} = \dfrac{\partial N}{\partial H_0} = 0\right)$ it can be shown that the relative change of E_F with the field is of the order of $(\hbar\omega_0/E_F)^2$ which is a very small quantity.

Thus the magnetization is given by the relation

$$\langle S_z \rangle = \langle S_z \rangle_0 (1-s),$$

and this relation is valid as long as the magnetization remains proportional to the applied field, that is to say $\hbar\omega_0 \ll k_B T_F$ (and not $\hbar\omega_0 \ll kT$ as in a paramagnetic system with localized spins). The two definitions of s are thus equivalent.

The Overhauser effect may be demonstrated by evaluating the rate of change of the nuclear populations in the presence of electronic saturation. For simplicity, a value $I = \frac{1}{2}$ is again assumed.

In Chapter 1 for a nuclear equilibrium situation we derived the equation

$$\frac{N_+}{N_-} = \frac{W_\uparrow}{W_\downarrow},$$

the rate $W_\uparrow$ being deduced from eqn (4.23); similarly $W_\downarrow$ has the value

$$W_\downarrow = \sum_{\mathbf{k},\mathbf{k'}} W_{\mathbf{k}\downarrow,\mathbf{k'}\uparrow}\, n(E_{\mathbf{k}\downarrow})\{1-n(E_{\mathbf{k'}\uparrow})\}.$$

With

$$W_{\mathbf{k}\downarrow,\mathbf{k'}\uparrow} = \frac{2\pi}{\hbar}\left(\frac{8\pi}{3}\gamma_n g\beta\hbar\right)^2 \tfrac{1}{4}|U_{\mathbf{k}}(\mathbf{R}_i)|^2|U_{\mathbf{k'}}(\mathbf{R}_i)|^2 \delta(E_{\mathbf{k}\downarrow}-E_{\mathbf{k'}\uparrow}+\hbar\omega_n) \quad (4.35)$$

the ratio of the populations becomes

$$\frac{N_+}{N_-} = \frac{\sum'_{\mathbf{k},\mathbf{k'}}|U_{\mathbf{k}}(\mathbf{R}_i)|^2|U_{\mathbf{k'}}(\mathbf{R}_i)|^2 n_\uparrow\{E_{\mathbf{k'}}-\hbar(\omega_0-\omega_n)\}\{1-n_\downarrow(E_{\mathbf{k'}})\}}{\sum'_{\mathbf{k},\mathbf{k'}}|U_{\mathbf{k}}(\mathbf{R}_i)|^2|U_{\mathbf{k'}}(\mathbf{R}_i)|^2 n_\downarrow(E_{\mathbf{k}})[1-n_\uparrow\{E_{\mathbf{k}}-\hbar(\omega_0-\omega_n)\}]}, \quad (4.36)$$

where the prime over the summation indicates that the summation over $\mathbf{k}$ and $\mathbf{k'}$ is restricted to those states that satisfy the conservation of energy conditions in eqns (4.35) and (4.23).

If the following ratio ρ is calculated,

$$\rho = \frac{n_\uparrow\{E-\hbar(\omega_0-\omega_n)\}\{1-n_\downarrow(E)\}}{n_\downarrow(E)[1-n_\uparrow\{E-\hbar(\omega_0-\omega_n)\}]},$$

we find

$$\rho = \exp\left(\frac{E_F^+ - E_F^- - \hbar\omega_n}{k_B T}\right),$$

and the remarkable fact is that ρ does not depend on the energy E.

Going back to eqn (4.36), the ratio of the nuclear populations is obtained:

$$\frac{N_+}{N_-} = \rho = \exp\left(\frac{E_{\mathrm{F}}^+ - E_{\mathrm{F}}^- - \hbar\omega_{\mathrm{n}}}{k_{\mathrm{B}}\,T}\right). \tag{4.37}$$

If the electronic system is not perturbed $E_{\mathrm{F}}^+ = E_{\mathrm{F}}^-$, and the ratio is given as expected by the Boltzmann factor. In the presence of an electronic saturation, using the relation (4.33) we obtain

$$\frac{N_+}{N_-} = \exp\left\{\frac{\hbar(s\omega_0 - \omega_{\mathrm{n}})}{k_{\mathrm{B}}\,T}\right\}, \tag{4.38}$$

which is eqn (4.30) with $\xi = -1$.

It is important to notice that although the Overhauser effect is not restricted to metals, the situation in metals differs greatly from the situation met in other systems like liquids. In a liquid containing paramagnetic ions the Overhauser effect has for result to enhance the nuclear polarization to a value equal (or comparable) to the electronic polarization. This affirmation in a metal does contradict eqn (4.38), because here the electronic polarization is strongly reduced by the Fermi–Dirac statistics. In a metal a complete Overhauser effect produces a nuclear polarization equal to the electronic polarization for electrons obeying Boltzmann statistics. Here again, it can be shown that the effect is not limited to nuclei having a spin value of $\frac{1}{2}$.

3.5. *Influence of the electron–electron interactions on the relaxation rate*

The relaxation rate due to the contact coupling may be written in a different form, using the N-electron energy levels and wave functions. The probability is given by the golden rule ($I = \frac{1}{2}$ is again assumed and also $W_\uparrow = W_\downarrow$)

$$\frac{1}{T_1} = \frac{4\pi}{\hbar}\left(\frac{8\pi}{3}\,\gamma_{\mathrm{n}}g\beta\hbar\right)^2 \sum_{E_{\mathrm{i}},\,E_{\mathrm{f}}} P_i\left|\left\langle E_{\mathrm{i}}\left|\frac{S_+(\mathbf{R}_i)}{2}\right|E_{\mathrm{f}}\right\rangle\right|^2 \delta(E_{\mathrm{i}} - E_{\mathrm{f}} - \hbar\omega_{\mathrm{n}}), \tag{4.39}$$

where P_i is the statistical weight for the N-electron state of energy and E_{i}, E_{f} is the energy of the final electronic state. This equation is more general than eqns (4.23)-(4.25) because no assumption has been made about the structure of the electronic states. (This equation is also valid for a superconductor.)

If the δ-function is replaced by its Fourier transform, the equation becomes

$$\frac{1}{T_1} = \frac{32\pi^2}{9}\,\gamma_{\mathrm{n}}^2 g^2\beta^2 \sum_{E_{\mathrm{i}},\,E_{\mathrm{f}}} \int_{-\infty}^{+\infty} \exp\left\{\frac{it}{\hbar}(E_{\mathrm{i}} - E_{\mathrm{f}} - \hbar\omega_{\mathrm{n}})\right\} \mathrm{d}t \times$$
$$\times\; P_i\langle E_{\mathrm{i}}|S_+(\mathbf{R}_i)|E_{\mathrm{f}}\rangle\langle E_{\mathrm{f}}|S_-(\mathbf{R}_i)|E_{\mathrm{i}}\rangle.$$

The quantity to be integrated may be transformed as follows:

$$\sum_{E_i, E_f} e^{-i\omega_{nl} t} P_i \langle E_i | e^{iE_i t/\hbar} S_+(\mathbf{R}_i) e^{-iE_f t/\hbar} | E_f \rangle \langle E_f | S_-(\mathbf{R}_i) | E_i \rangle,$$

or

$$e^{-i\omega_{nl} t} \operatorname{Tr}\{P S_+(\mathbf{R}_i, t) S_-(\mathbf{R}_i, 0)\},$$

with the definition

$$S_+(\mathbf{R}_i, t) = e^{i\mathscr{H}_{el} t/\hbar} S_+(\mathbf{R}_i) e^{-i\mathscr{H}_{el} t/\hbar},$$

$\mathscr{H}_e$ being the total electronic Hamiltonian and P the density matrix of the electronic system.

The final result is

$$\frac{1}{T_1} = \frac{32\pi^2}{9} \gamma_n^2 g^2 \beta^2 \int_{-\infty}^{+\infty} e^{-i\omega_{nl} t} \, dt \, \langle S_+(\mathbf{R}_i, t) S_-(\mathbf{R}_i, 0) \rangle, \qquad (4.40)$$

where the average symbol means a thermal average over the states of the electronic system. The relaxation time is thus expressed in terms of a correlation function. And as a correlation function can be expressed in term of a generalized susceptibility (see Ref. 22, p. 400), this equation will allow us, at least in a formal way, to take the interactions into account. This point of view is presented in an article by P. A. Wolff.[23]

Following References 22 and 23, it can be shown that $1/T_1$ may be written (more details are found in Appendix 1)

$$\frac{1}{T_1} \sim \frac{ik_B T}{\omega_n} \{\chi_S(\mathbf{R}_i, \mathbf{R}_i, \omega_n) - \chi_S(\mathbf{R}_i, \mathbf{R}_i, -\omega_n)\},$$

or, using the Fourier transform of $\chi_S(\mathbf{R}_i, \mathbf{R}_i)$,

$$\frac{1}{T_1} \sim \frac{ik_B T}{\omega_n} \sum_{\mathbf{q}} \{\chi_S(\mathbf{q}, \omega_n) - \chi_S(\mathbf{q}, -\omega_n)\}.$$

If now we take for $\chi_S(\mathbf{q}, \omega_n)$ the corrected value given by (3.32) it is possible to estimate the effect of the exchange interaction on $1/T_1$. Equation (3.32) was presented for $\chi_S(\mathbf{q}, 0)$ but as ω_n remains extremely small, this equation will be still valid; the following result is obtained:

$$\frac{1}{T_1} \sim \frac{ik_B T}{\omega_n} \sum_{\mathbf{q}} \frac{\chi_S^F(\mathbf{q}, \omega_n) - \chi_S^F(\mathbf{q}, -\omega_n)}{[1 + B_0 \{\chi_S^F(\mathbf{q})/\chi_S^F\}]^2}.$$

As the quantity $\chi_S^F(\mathbf{q}, \omega_n) - \chi_S^F(\mathbf{q}, -\omega_n)$ may be expressed in terms of a correlation function of $S_\mathbf{q}^z S_{-\mathbf{q}}^z$ for a free-electron gas, knowing B_0 the influence of the correlation may be calculated by numerical integration. This calculation is rather uncertain because the model of Wolff is presumably wrong when $|\mathbf{q}|$ is large. On using these equations it is found that the relaxation rate is increased by about 20 per cent for metals such as sodium and lithium.

There is a problem which, as far as we know, has not been solved. It concerns the calculation of the Overhauser effect in an interacting electron gas. The calculation we have described is not valid. It is certainly not correct for an interacting electron gas to introduce two energy bands with equal or different Fermi energies.

In Chapter 3 we showed that the energy required for changing the orientation of the spin of a quasi-particle is not $\hbar\omega_0$ but $\hbar\omega_0/(1+B_0)$, where B_0 is defined by eqn (3.23). If for the Overhauser effect a calculation similar to the one we described earlier is tried, we shall obtain a polarization that differs from the polarization given by the Boltzmann factor involving the electronic Zeeman energy. This result violates a general thermodynamic description of the Overhauser effect.[24] The explanation of this contradiction lies in the fact that when the electronic spin system is saturated, we are not changing the orientation of one quasi-particle in the exchange field due to the others but changing the spin orientation of all the quasi-particles, i.e. one is exciting a collective mode. This point will appear more clearly in the discussions of the last chapter.

3.6. *The Overhauser shift*

Instead of measuring the Overhauser enhancement of the nuclear polarization by a measurement of a change in the intensity of the nuclear signal, it is also possible to measure the shift of the electronic line itself. This method is convenient for experimental reasons since all the measurements are performed without observing a nuclear signal. In the high-temperature region, using eqns (4.32) and (4.8), we obtain for the nuclear field

$$\langle H_{nz} \rangle = -\frac{8\pi}{3}\hbar\gamma_n\langle I_z\rangle_0|U_{k_F}(\mathbf{R}_i)|^2\left(1+s\xi\frac{\gamma_e}{\gamma_n}\right). \qquad (4.41)$$

Because the ratio γ_e/γ_n is large this field may be very important. This effect was observed for sodium and lithium and the results are discussed in Chapter 6.

3.7. *Other possible contributions to the nuclear relaxation rate in metals*

We have discussed in great detail the relaxation processes coming from the magnetic part of the hyperfine interaction but in some cases other mechanisms are also present. There is a relaxation process that is very efficient at temperatures where the ions are diffusing. The diffusion of the ions produces a modulation of the spin–spin interaction that is the source of a relaxation process. In Chapter 5 it will be shown that the quadrupolar hyperfine interaction is also sometimes a source of relaxation.

4. The shape of the nuclear resonance line and the indirect spin–spin interactions

4.1. *Introduction to the calculation of the line shape of a nuclear resonance line*

This problem was briefly introduced in the first chapter where the discussion was restricted to the influence on the line shape of the classical dipolar interaction between nuclear moments. In metals (and particularly in heavy metals) other interactions between nuclear spins are important; they are called indirect interactions. In a rather general way these interactions may be written

$$\mathcal{H}_{\text{ind}} = \sum_{i,j,\alpha,\beta} A_{ij}^{\alpha\beta} I_i^\alpha I_j^\beta,$$

where $A_{ij}^{\alpha\beta}$ is a tensor. Its components are functions of the distance r_{ij} between the nuclei and also of the orientation of the unit vector $\mathbf{n}_{ij}$ directed along the line joining the two nuclei with respect to the crystalline axis.

This interaction takes simpler form in the following cases. First, if the interaction is calculated using only the scalar hyperfine coupling it will be shown in the next section that the indirect interaction is scalar and will be written

$$\mathcal{H}_{\text{ind}} = \sum_{i,j} A_{ij} \mathbf{I}_i \cdot \mathbf{I}_j.$$

On the other hand, if other terms in the hyperfine coupling are considered but if the Fermi surface is spherical it can be shown[25] that the interaction has a scalar part and another part of the following form:

$$\mathcal{H}_{\text{ind}} = \sum_{i,j} B_{ij} \{ (\mathbf{I}_i \cdot \mathbf{n}_{ij})(\mathbf{I}_j \cdot \mathbf{n}_{ij}) - \tfrac{1}{3} \mathbf{I}_i \cdot \mathbf{I}_j \}.$$

This interaction is called the pseudo-dipolar interaction.

In other situations the only result is that the tensor $A_{ij}^{\alpha\beta}$ is symmetrical and therefore can be diagonalized and further split into a scalar and a traceless part.

In all metals where indirect interactions are observed the scalar part is the largest, and for simplicity we shall assume that the interaction is the sum of a large scalar interaction and a small pseudo-dipolar interaction.

As mentioned in Chapter 1, it is not possible to predict the line shape exactly and so the moments method will be used. The results of this method are given without proof (Ref. 1, chap. 4).

(1) If the Zeeman energy is large compared to all the spin–spin interactions, only those terms of the Hamiltonian of the spin–spin interaction that commute with the Zeeman Hamiltonian have to be considered. (We assume that only one type of nuclear moment is present in the metal.) There are two such terms: the scalar coupling and in the dipolar or pseudo-dipolar coupling, the term $I_{iz}I_{jz}-\frac{1}{3}\mathbf{I}_i.\mathbf{I}_j$. This new spin–spin Hamiltonian is called the truncated Hamiltonian and will be written $\mathscr{H}'_1$.

(2) The second moment of the line is given by the equation

$$M_2 = \frac{1}{\hbar^2}\frac{\mathrm{Tr}[\mathscr{H}'_1, I_x]^2}{\mathrm{Tr}[I_x^2]}, \tag{4.42}$$

with $I_x = \sum_i I_{ix}$. It is sometimes useful to calculate the fourth moment M_4,

$$M_4 = \frac{1}{\hbar^4}\frac{\mathrm{Tr}\{[\mathscr{H}'_1,[\mathscr{H}'_1, I_x]]\}^2}{\mathrm{Tr}[I_x^2]}. \tag{4.43}$$

These moments may be calculated without any approximation. (The calculation of higher-order moments, although possible in principle, will necessitate very complicated algebraic calculations and is very seldom done.)

The problem remains to predict the line shape knowing only these two moments. Mathematically speaking, this problem has no solution because the line shape is known only if all the moments are known. The standard and physical way of solving this problem is to try simple line shape. First a comparison is made between M_2^2 and M_4. For a Gaussian line shape

$$\chi''(\omega) \simeq \exp\left(-\frac{(\omega-\omega_n)^2}{2\Delta^2}\right)$$

it can be verified that $M_4 = 3M_2^2$. If the calculated M_2^2 (using eqn (4.42)) does not differ too much from M_4 (4.43), a Gaussian line shape will be a reasonable approximation of the real line shape. For a Gaussian line the line width is proportional to $M_2^{\frac{1}{2}}$ and again if M_4 is found to be comparable to M_2^2 we will take $M_2^{\frac{1}{2}}$ as an estimate of the line width. For an interaction that has scalar and dipolar components, the second moment, using eqn (4.42), has the value

$$\hbar^2 M_2 = \frac{I(I+1)}{3}\sum_j B_{ij}^2. \tag{4.44}$$

(The classical dipolar interaction is included in the definition of B_{ij}.) The important result is that this moment does not depend upon the magnitude of the scalar coupling.

The measurement of the line width in this situation ($M_2^2 \simeq M_4$) does not permit a measurement of the amplitude of the scalar coupling. In other situations the calculation leads to a value of M_4 much larger than M_2^2. As an example, such a result is obtained when the scalar coupling is large compared to the dipolar (and pseudo-dipolar) interactions because the scalar coupling does not contribute to M_2 but does contribute to M_4, the largest term being of the order of

$$M_4 \sim \sum_{j,k} B_{ij}^2 A_{ik}^2.$$

The line shape cannot be Gaussian. A Lorentzian shape is usually assumed, of the form

$$\chi''(\omega) \simeq \frac{1}{\Delta^2 + (\omega - \omega_n)^2}.$$

A true Lorentzian shape cannot be a reasonable choice because this shape leads to infinite values for all the moments. Thus we assume a truncated Lorentzian shape; the line is Lorentzian only when $|\omega - \omega_n| < \alpha$, where $\alpha \gg \Delta$. When $|\omega - \omega_n| > \alpha$, $\chi''(\omega) = 0$. With this assumption α and Δ are evaluated as functions of M_2 and M_4, and the line width Δ is given by the relation (Ref. 1, p. 108)

$$\Delta = \frac{\pi}{2\sqrt{3}} M_2^{\frac{1}{2}} \left(\frac{M_2^2}{M_4}\right)^{\frac{1}{2}}. \tag{4.45}$$

The numerical coefficient in front of this equation has no physical significance. Its value may be changed if the form of the truncation of the Lorentzian shape is changed. However, the order of magnitude of the line width Δ will certainly be given by eqn (4.45). By using a completely different method, Anderson and Weiss[26] have obtained a similar equation. We notice that now the line width Δ is much smaller than the square root of the second moment. This effect is called 'the exchange narrowing'.

(3) Another situation met in metals is when two isotopes with different gyromagnetic ratios are present (the same analysis will be valid in an alloy with two different elements). It is possible to study the line shape of the two resonance signals. Let us call I and S the spins of the two species. The spin–spin interaction Hamiltonian can be separated into three parts: $\mathscr{H}_{II}$ for the interactions between the spins I, $\mathscr{H}_{SS}$ for the interactions among the S spins, and $\mathscr{H}_{IS}$ for the interactions between I and S spins. Let us give the rules to calculate the moments of the resonance line for the spins I. Again only the parts of the interaction Hamiltonian that commute with the operator I_z will be considered. We

find terms coming from $\mathscr{H}_{II}$, defined as before and called $\mathscr{H}'_{II}$, all the terms of $\mathscr{H}_{SS}$, and finally, in $\mathscr{H}_{IS}$, only terms like $I_{iz} S_{i'z}$, which are called $\mathscr{H}'_{IS}$.

The moments are still given by eqns (4.42) and (4.43), provided we use for $\mathscr{H}'_1$ the following three terms:

$$\mathscr{H}'_1 = \mathscr{H}'_{II} + \mathscr{H}_{SS} + \mathscr{H}'_{IS}.$$

In the second moment calculation $\mathscr{H}_{SS}$ does not contribute, $\mathscr{H}'_{II}$ gives the contribution shown in eqn (4.44), and $\mathscr{H}'_{IS}$ also contributes. A term like $I_{iz} S_{i'z}$ in $\mathscr{H}'_{IS}$ has contributions from both the dipolar and scalar interactions. Therefore a scalar coupling between unlike spins contributes to the second moment and thus to the broadening of the line. The calculation of the fourth moment is rather tedious[27] as the value also depends on the $\mathscr{H}_{SS}$ Hamiltonian. It is not possible to make general predictions about the value of the ratio M_4/M_2^2 for this number is a function of too many parameters, namely the relative abundance of the two species, the ratio of their gyromagnetic ratios, and the relative magnitude of the scalar and dipolar interactions. Some examples will be discussed and compared with the experimental results in Chapter 6.

4.2. *Calculation of indirect interactions*

The physical origin of indirect interactions is as follows. The nuclear moment at the point $\mathbf{R}_i$ creates a local magnetic perturbation which induces an electronic magnetization varying in space which in its turn interacts with another nuclear moment at the point $\mathbf{R}_j$. The net result of this effect may be described by a static coupling between the nuclear moments. From this description we notice that this interaction, when the coupling is via the contact hyperfine term, is simply proportional to the susceptibility $\chi_S(\mathbf{R}_j - \mathbf{R}_i)$ defined in Chapter 3. The first calculation performed by Ruderman and Kittel[28] was done using second-order perturbation theory and neglecting the electron–electron interactions. Let us first consider the effect of the hyperfine contact interactions of the two nuclei. The energy calculated by a second-order perturbation method is given by the equation

$$\mathscr{H}_{ij} = \left(\frac{8\pi}{3} g\beta\hbar\gamma_n\right)^2 \sum_{E_i, E_f} P_i \frac{\langle E_i | \mathbf{I}_i . \mathbf{S}(\mathbf{R}_i) | E_f \rangle \langle E_f | \mathbf{I}_j . \mathbf{S}(\mathbf{R}_j) | E_i \rangle}{E_f - E_i} +$$

$$+ P_i \frac{\langle E_i | \mathbf{I}_j . \mathbf{S}(\mathbf{R}_j) | E_f \rangle \langle E_f | \mathbf{I}_i . \mathbf{S}(\mathbf{R}_i) | E_i \rangle}{E_f - E_i}, \quad (4.46)$$

where we use the notation already defined for the relaxation time calculation (eqn (4.37)). There is indeed a similarity between the two calculations which illustrates the fact already mentioned that a correlation function may be expressed as a function of the non-local susceptibilities. The calculation may be simplified by assuming a low temperature, thus taking for E_i the ground state E_0 (such an approximation was of course not made for the relaxation rate because this rate vanishes when T tends toward zero).

For non-interacting electrons eqn (4.46) becomes

$$\mathscr{H}_{ij} = \left(\frac{8\pi}{3} g\beta\hbar\gamma_n\right)^2 \times$$

$$\times \sum_{\substack{E_{k\sigma} < E_F \\ E_{k'\sigma'} > E_F}} \frac{\langle \Phi_{k\sigma} | \mathbf{I}_i . \mathbf{s}\delta(\mathbf{r}-\mathbf{R}_i) | \Phi_{k'\sigma'} \rangle \langle \Phi_{k'\sigma'} | \mathbf{I}_j . \mathbf{s}\delta(\mathbf{r}-\mathbf{R}_j) | \Phi_{k\sigma} \rangle}{E_{k'\sigma'} - E_{k\sigma}} +$$

$$+\text{complex conjugate}$$

if the nuclear Zeeman energy is neglected. If the electronic Zeeman energy is also neglected a scalar interaction between the nuclear moments is found:

$$\mathscr{H}_{ij} = \tfrac{1}{2}\mathbf{I}_i . \mathbf{I}_j \left(\frac{8\pi}{3} g\beta\hbar\gamma_n\right)^2 \sum_{\substack{E_k < E_F \\ E_{k'} > E_F}} \frac{|U_k(\mathbf{R}_i)|^2 |U_{k'}(\mathbf{R}_i)|^2 e^{i\mathbf{R}_{ij}.(\mathbf{k}-\mathbf{k}')}}{E_k - E_{k'}} +$$

$$+\text{complex conjugate.} \quad (4.47)$$

The largest contribution to the summation comes when the two energies are comparable and consequently both in the vicinity of the Fermi energy E_F. We shall replace the squares of the wave functions $|U_k(\mathbf{R}_i)|^2$ by their average value at the Fermi surface. A similar approximation was used for the evaluation of the Knight shift and of the relaxation rate but in those two cases the approximation was justified by the occurrence in the summation of a delta function $\delta(E_k - E_F)$; here the justification of the approximation is somewhat weaker.

With this approximation eqn (4.46) may be written as

$$\mathscr{H}_{ij} = \tfrac{1}{2}\mathbf{I}_i . \mathbf{I}_j \left(\frac{8\pi}{3} \hbar\gamma_n\right)^2 |U_{k_F}(\mathbf{R}_i)|^4 \chi_S(\mathbf{R}_i-\mathbf{R}_j), \quad (4.48)$$

where $\chi_S(\mathbf{R}_i-\mathbf{R}_j)$ is the non-local susceptibility for a free-electron gas in the absence of the lattice. The Fourier transform of $\chi_S(\mathbf{R}_i-\mathbf{R}_j)$ is the function $\chi_S(\mathbf{q})$ given by eqn (3.30) and for a parabolic and isotropic energy band by eqn (3.31). If we calculate the Fourier transform, the following result for the interaction is obtained:

$$\mathscr{H}_{ij} = m^* \mathbf{I}_i . \mathbf{I}_j \left(\frac{8\pi}{3} \hbar g\beta\gamma_n\right)^2 \frac{|U_{k_F}(\mathbf{R}_i)|^4}{4\hbar(2\pi)^3} F(R_{ij}), \quad (4.49)$$

where the function $F(R)$ has the value

$$F(R) = \frac{1}{R^4}\{2k_F R \cos(2k_F R) - \sin(2k_F R)\}. \qquad (4.50)$$

The function $F(R)$ is presented on Fig. 4.2. The interaction decreases if the distance between the nuclei is increased presenting characteristic oscillations. As for the nuclear relaxation rate, more sophisticated

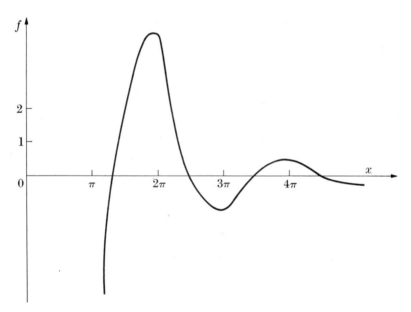

FIG. 4.2. Variation of the function $f = 10^3(x \cos x - \sin x)/x^4$.

calculations are possible using corrected values for the non-local susceptibility in eqn (4.48). Experimentally this interaction is accessible only for distances corresponding to nearest neighbours and the value of $\chi_S(\mathbf{R})$ comes from the behaviour of $\chi_S(\mathbf{q})$ for large values of $\mathbf{q}$, a region where $\chi_S(\mathbf{q})$ is not accurately known.

4.3. Non-scalar indirect interactions

Other indirect interactions are found if the other terms of the hyperfine interactions are considered. The calculation is performed using the same technique. First, there are the contributions involving the orbital term for one nuclear moment and the contact or dipolar term for the other, but these contributions are very small because they involve the average value of the electronic spin operator. They are therefore reduced by a factor $g\beta H_0/E_F$ compared to the others.

The largest contribution occurs if one considers the contact interaction for one nuclear moment and the dipolar interaction for the other. The calculation will again involve matrix elements like the $\mathscr{T}_{kk'}^{\alpha\beta}$ defined by eqn. (4.11). The final result has the form of a pseudo-dipolar coupling between the nuclear moments[25] if the Fermi surface is a sphere.

It will be seen later that this term can have a value as large as 30 per cent of the scalar indirect interaction.

4.4. *Nuclear line shape in the presence of indirect interactions*

The largest part of the indirect interactions is always the scalar interaction. We shall discuss some of the situations met in metals, when the indirect interactions become comparable or larger than the classical dipolar interaction.

4.4.1. *One type of nuclear moment.* The scalar indirect interaction does not contribute to the second moment. Therefore when the indirect interactions are of a magnitude comparable to the classical dipolar interactions, they are very difficult to measure by a study of the line shape.

If the indirect interactions become large an exchange narrowing will appear as the scalar part always predominates. This effect is observed for the nuclear resonance in caesium and platinum, a case that will be discussed at more length in Chapter 6.

4.4.2. *Two types of nuclear moment.* Here we have to remember that a scalar coupling between unlike spins produces a broadening. It is now possible to measure the indirect interaction even if it is of the same order of magnitude as the classical dipolar coupling, by measuring the line width. This situation occurs in the nuclear resonance of the two silver isotopes [107]Ag and [109]Ag. This example is of historical interest because it was the observation of a broadening for the [109]Ag resonance line that prompted the calculation of Ruderman and Kittel. Some other examples of this behaviour will be discussed later.

4.5. *Other sources of broadening*

For the sake of completeness we mention that the quadrupolar effects very often lead to a broadening of the line, but we leave the discussion of this question until Chapters 5 and 7.

As the resonance signal is almost always observed when using poly-crystalline samples, the existence of an anisotropic Knight shift is also a source of broadening. This broadening is easily identified because it is proportional to the external field. The effect gives rise to a characteristic line shape which we shall now analyse.

Let us assume the existence of an anisotropic shift for a nuclear moment in a site having a quadratic symmetry (in eqn (4.16) this symmetry implies $\eta_d = 0$). The resonance frequency is given by the equation

$$\nu = \nu_0 + \tfrac{1}{2}\alpha(3\cos^2\theta - 1)$$

(the isotropic shift is included in ν_0), where θ is the angle between the quadratic axis and the magnetic field and α is proportional to the value of the anisotropic shift.

In a powdered sample all the possible values for the quantity $u = \cos\theta$ are equally probable. The intensity of the signal as a function of ν will be called $g(\nu)$. The function $g(\nu)$ is defined so that $g(\nu)\,d\nu$ is the number of nuclear spins having a resonance frequency between the values ν and $\nu + d\nu$ (for a given value of the external field). Taking into account the equal probability for all the values of the variable u the function $g(\nu)$ is defined by the equation

$$g(\nu)\,d\nu = du,$$

or
$$g(\nu) = 1 \Big/ \frac{d\nu}{du} \propto \frac{1}{\alpha u} = \frac{1}{\sqrt{\{2(\nu-\nu_0)+\alpha\}}}, \tag{4.51}$$

with the condition $\qquad \alpha < (\nu-\nu_0) < -\tfrac{1}{2}\alpha.$

The line shape is presented on Fig. 4.3. Equation (4.51) predicts an infinite value for $g(\nu)$ when $\nu = \nu_0 - \tfrac{1}{2}\alpha$. This behaviour is not observed because other sources of broadening always exist which suppress this singularity. The line shape is very asymmetrical and this fact, together with the variation with the external field, is a characteristic of this type of broadening.[29]

4.6. Line shape and diffusion effects

Throughout the discussion about the line shape, the spin–spin interactions were assumed to be static. In metals well below the melting point the diffusion motion will produce a time variation of the interactions (see Ref. 1, chap. 10). If the motion is fast enough, only the average value of the interaction will contribute to the broadening. For dipolar or pseudo-dipolar interactions the averaged value is zero because of the angular dependence. When the diffusion process appears, a narrowing of the line is observed. This effect was experimentally studied in detail for light metals where there is no indirect interaction. For heavy metals at high temperature the line width is due to the relaxation rate processes and the observation of the diffusion narrowing is difficult.

5. Relaxation in weak magnetic fields: the spin temperature concept

In all discussions so far it has been assumed that the spin–spin interactions were small compared to the nuclear Zeeman energy. We shall consider now situations where this condition is no longer fulfilled. This

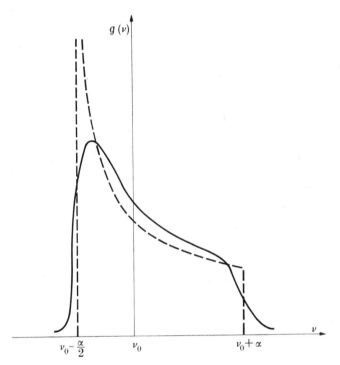

Fig. 4.3. The theoretical line shape due to an anisotropic Knight shift in a powder (broken curve). The full curve gives the line shape if another source of broadening is taken into account.

problem is very interesting in itself from the point of view of the theory of nuclear magnetism. From our point of view, the quantities that are introduced using the spin-temperature theory provide new information about the metallic properties. We will deal with only the basic assumptions of these theories; more details may be found in Reference 1, chap. 12 and Reference 30.

5.1. *The spin temperature assumption*

When the spin–spin interactions become important, it is no longer possible to consider the energy levels for a given nuclear spin; one must deal with the system of all the nuclear moments. The usual approach uses the concept of a density matrix. For a nuclear-spin system in an equilibrium situation at a temperature T the density matrix ρ is given by the formula

$$\rho = e^{-\mathcal{H}/k_B T}/\mathrm{Tr}(e^{-\mathcal{H}/k_B T}), \tag{4.52}$$

where $\mathcal{H}$ is the total nuclear Hamiltonian

$$\mathcal{H} = \mathcal{H}_0 + \mathcal{H}_I = -\hbar\gamma_n H_0 \sum_i I_{iz} + \mathcal{H}_I,$$

$\mathcal{H}_I$ being the Hamiltonian of the spin–spin interactions. For simplicity we consider a system with only one type of nuclear moment. The generalization to include other situations is quite straightforward.

It is usually sufficient to use only the first two terms in the expansion of ρ as a function of the inverse of the temperature:

$$\rho \simeq 1 - \frac{\mathcal{H}}{k_B T}. \tag{4.53}$$

The spin temperature assumption is expressed by the following statement. For some situations, the behaviour of the nuclear-spin system can be described by a density matrix ρ keeping the form given by eqns (4.52) and (4.53), but with a value of T that will be written T_S and may be equal to or different from the temperature of the other degrees of freedom.

Using this assumption it is possible to extend the definition of the relaxation time T_1. We will now define T_1 as the time constant for the evolution of the inverse spin temperature $1/T_S$ towards its equilibrium value $1/T$:

$$\frac{\mathrm{d}}{\mathrm{d}t}\left(\frac{1}{T_S}\right) = -\frac{1}{T_1}\left(\frac{1}{T_S} - \frac{1}{T}\right). \tag{4.54}$$

(We have assumed that the evolution of $1/T_S$ is exponential.)

This new definition is in agreement with the definition of T_1 given in Chapter 1 (eqn (1.12)) when the spin–spin interactions are neglected. Let us demonstrate this property. The average value of the magnetization m_z is expressed using the density matrix by the equation

$$m_z = \mathrm{Tr}\left\{\hbar\gamma_n \sum_i (I_{iz})\rho\right\};$$

if $\mathcal{H}_I$ is neglected, using eqn (4.53), it is found that

$$m_z \simeq \frac{(\gamma_n \hbar)^2}{k_B T_S} H_0 \, \mathrm{Tr}\left(\sum_i I_{iz}\right)^2,$$

i.e. the evolution of m_z will be proportional to the evolution of $1/T_S$. This new definition of T_1 may also be used to extend our calculations to situations where I is larger than $\frac{1}{2}$ (however it is not at all necessary to make the spin temperature assumption for extending the relaxation rate equation (4.26) for $I > \frac{1}{2}$ in a high magnetic field; it can be shown that for a magnetic relaxation process, which is the situation met in metals, eqns (4.26) or (4.40) are valid if $I > \frac{1}{2}$ without using the spin temperature assumption (see Ref. 1 and Ref. 3, p. 125)).

For calculating the relaxation rate the evolution of the density matrix ρ as a function of time (due to the coupling between the nuclear spins and the lattice) is evaluated by a perturbation approach. The details of the calculations will not be given here (see Ref. 1, pp. 361 and 362 and Ref. 30). We find that eqn (4.54) is valid with T_1 given by the equation

$$\frac{1}{T_1} = \frac{1}{\mathrm{Tr}(\mathscr{H}^2)} \int\limits_{-\infty}^{+\infty} d\tau\, \mathrm{Tr}(\langle [\mathscr{H}'(\tau), \mathscr{H}][\mathscr{H}'(0), \mathscr{H}]\rangle) \qquad (4.55)$$

with the following definitions:

$$\mathscr{H}'(\tau) = \exp\!\left(\frac{i}{\hbar}(\mathscr{H}+\mathscr{F})\tau\right)\mathscr{H}' \exp\!\left(-\frac{i}{\hbar}(\mathscr{H}+\mathscr{F})\tau\right).$$

$\mathscr{H}'$ is the Hamiltonian describing the coupling of the nuclear spins with the lattice (here the hyperfine interaction) and $\mathscr{F}$ is the 'lattice' Hamiltonian. The symbol $\langle\ \rangle$ means that an average thermal value is taken over the lattice variables and finally the trace is taken over the nuclear variables. Although eqn (4.55) looks complicated, if for $\mathscr{H}'$ we take the contact hyperfine interaction and neglect $\mathscr{H}_I$ in $\mathscr{H}$, the equation reduces exactly to eqn (4.40) after taking the trace over the nuclear variables.

The relaxation time T_1 defined by eqn (4.55) will depend on the value of the external magnetic field because $\mathscr{H}$ is a function of H_0, and this T_1 compared to the value obtained in the very high magnetic field that we calculated earlier, which will be written $T_1(\infty)$, is given by

$$\frac{1}{T_1} = \frac{1}{T_1(\infty)} \frac{\mathrm{Tr}(\mathscr{H}_0^2)}{\mathrm{Tr}(\mathscr{H}_0+\mathscr{H}_I)^2} \frac{\int\limits_{-\infty}^{+\infty} d\tau\, \mathrm{Tr}(\langle [\mathscr{H}'(\tau), \mathscr{H}_0+\mathscr{H}_I][\mathscr{H}'(0), \mathscr{H}_0+\mathscr{H}_I]\rangle)}{\int\limits_{-\infty}^{+\infty} d\tau\, \mathrm{Tr}(\langle [\mathscr{H}'(\tau), \mathscr{H}_0][\mathscr{H}'(0), \mathscr{H}_0]\rangle)}.$$

$$(4.56)$$

This equation is obtained by calculating $T_1(\infty)$ using eqn (4.55) in which the terms involving $\mathscr{H}_1$ are neglected. In the trace calculations the cross terms involving products $\mathscr{H}_0\mathscr{H}_I$ disappear, and the result can be written

$$\frac{1}{T_1} = \frac{1}{T_1(\infty)} \frac{H_0^2+\delta H_{\mathrm{L}}^2}{H_0^2+H_{\mathrm{L}}^2} \qquad (4.57)$$

with the following definitions:

$$\hbar^2\gamma^2 H_{\mathrm{L}}^2 = \mathrm{Tr}(\mathscr{H}_I)^2, \qquad (4.58)$$

H_{L} is called the local field, and

$$\delta = \frac{H_0^2}{H_{\mathrm{L}}^2} \frac{\displaystyle\int_{-\infty}^{+\infty} \mathrm{d}\tau\, \mathrm{Tr}(\langle[\mathscr{H}'(\tau),\, \mathscr{H}_I][\mathscr{H}'(0),\, \mathscr{H}_I]\rangle)}{\displaystyle\int_{-\infty}^{+\infty} \mathrm{d}\tau\, \mathrm{Tr}(\langle[\mathscr{H}'(\tau),\, \mathscr{H}_0][\mathscr{H}'(0),\, \mathscr{H}_0]\rangle)}. \tag{4.59}$$

Equation (4.57) has the advantage of showing explicitly the field dependence of the relaxation rate. H_{L} gives the magnitude of the spin–spin coupling. The quantity δ does not depend on the value of the magnetic field and is physically interpreted as the ratio between the relaxation rate for the spin–spin energy and the Zeeman energy.

The calculation of H_{L} knowing the spin–spin interactions is quite straightforward. We find

$$\hbar^2\gamma^2 H_{\mathrm{L}}^2 = \frac{5I(I+1)}{9} \sum_j B_{ij}^2 + I(I+1) \sum_j A_{ij}^2. \tag{4.60}$$

The calculation of δ in a metal will now be discussed.

Let us first consider only two nuclear spins at the points $\mathbf{R}_i$ and $\mathbf{R}_j$ coupled by a scalar spin–spin interaction

$$\mathscr{H}_I = A_{ij}\,\mathbf{I}_i.\mathbf{I}_j.$$

The spin lattice Hamiltonian will be the contact term

$$\mathscr{H}' = \frac{8\pi}{3} g\beta\hbar\gamma_{\mathrm{n}}\{\mathbf{S}(\mathbf{R}_i).\mathbf{I}_i + \mathbf{S}(\mathbf{R}_j).\mathbf{I}_j\};$$

neglecting the nuclear Zeeman energy, we obtain for $\mathscr{H}'(\tau)$,

$$\mathscr{H}'(\tau) = \frac{8\pi}{3} g\beta\hbar\gamma_{\mathrm{n}}\{\mathbf{S}(\mathbf{R}_i, \tau).\mathbf{I}_i + \mathbf{S}(\mathbf{R}_j, \tau).\mathbf{I}_j\}.$$

In the commutators of the numerator of eqn (4.59) two types of terms appear. One is of the form

$$\int_{-\infty}^{+\infty} \mathrm{d}\tau\, \mathrm{Tr}(\langle[\mathbf{S}(\mathbf{R}_i, \tau).\mathbf{I}_i,\, A_{ij}\,\mathbf{I}_i.\mathbf{I}_j][\mathbf{S}(\mathbf{R}_i, 0).\mathbf{I}_i,\, A_{ij}\,\mathbf{I}_i.\mathbf{I}_j]\rangle),$$

and after taking the trace will involve a correlation function of the electronic spin already used in the relaxation calculation (eqn (4.40)). We have, of course, a similar term with $\mathbf{S}(\mathbf{R}_j, \tau)$. The other term is

$$\int_{-\infty}^{+\infty} \mathrm{d}\tau\, \mathrm{Tr}(\langle[\mathbf{S}(\mathbf{R}_j, \tau).\mathbf{I}_j,\, A_{ij}\,\mathbf{I}_i.\mathbf{I}_j][\mathbf{S}(\mathbf{R}_i, 0).\mathbf{I}_i,\, A_{ij}\,\mathbf{I}_i.\mathbf{I}_j]\rangle)$$

and is a function of the spin densities at different points.

The detailed calculations of the commutators and traces are lengthy but straightforward. The following results are obtained, if we call δ_{ij} the coefficient δ, considering only the two spins $\delta_{ij} = 2 + \epsilon_{ij}$ for a dipolar coupling and $\delta_{ij} = 2 - 2\epsilon_{ij}$ for a scalar coupling. The coefficient ϵ_{ij} is a measure of the ratio of the two types of correlation functions that appear in the calculation

$$\epsilon_{ij} = \frac{\int\limits_{-\infty}^{+\infty} d\tau \, \langle S_+(\mathbf{R}_i, \tau) S_-(\mathbf{R}_j, 0) \rangle}{\int\limits_{-\infty}^{+\infty} d\tau \, \langle S_+(\mathbf{R}_i, \tau) S_-(\mathbf{R}_i, 0) \rangle}. \qquad (4.61)$$

If there is no correlation between the electronic spin fluctuations at different points, $\epsilon_{ij} = 0$, and the spin–spin energy relaxes twice as fast as the Zeeman energy. On the other hand, if the electronic magnetization has the same value at the two points, the scalar interaction that now commutes with the hyperfine interaction is not relaxed, $\epsilon_{ij} = 1$, hence $\delta_{ij} = 0$.

As for the relaxation rate, ϵ_{ij} can be expressed in terms of the non-local susceptibilities and there is a relation between the function $\epsilon(\mathbf{R}_{ij})$ and the function $F(\mathbf{R}_{ij})$ which is proportional to $\chi_S(\mathbf{R}_i - \mathbf{R}_j)$. The calculation of $\epsilon(\mathbf{R}_{ij})$ will be presented at the end of this chapter.

Knowing the value of ϵ_{ij} for all the pairs of nuclear spins, the value of δ in eqn (4.59) is found and thus the relaxation rate

$$\frac{1}{T_1} = \frac{1}{T_1(\infty)} \frac{H_0^2 + \sum\limits_{i<j}(2+\epsilon_{ij})H_{\mathrm{d}ij}^2 + \sum\limits_{i<j}(2-2\epsilon_{ij})H_{\mathrm{s}ij}^2}{H_0^2 + \sum\limits_{i<j} H_{\mathrm{d}ij}^2 + \sum\limits_{i<j} H_{\mathrm{s}ij}^2}, \qquad (4.62)$$

where $H_{\mathrm{d}ij}$ is a dipolar local field defined by the relation

$$\hbar^2 \gamma^2 H_{\mathrm{d}ij}^2 = \tfrac{5}{9} I(I+1) B_{ij}^2$$

and similarly $H_{\mathrm{s}ij}$ is given by

$$\hbar^2 \gamma^2 H_{\mathrm{s}ij}^2 = I(I+1) A_{ij}^2.$$

Or, if we use eqn (4.60) for the local field,

$$\frac{1}{T_1} = \frac{1}{T_1(\infty)} \frac{H_0^2 + 2H_{\mathrm{L}}^2 + \sum\limits_{i<j} \epsilon_{ij}(H_{\mathrm{d}ij}^2 - 2H_{\mathrm{s}ij}^2)}{H_0^2 + H_{\mathrm{L}}^2}. \qquad (4.63)$$

The function ϵ_{ij} decreases very fast as R_{ij} increases and the last term of the numerator is noticeable only if the two spins i and j are nearest neighbours. If $\epsilon_{ij} = 0$ even for nearest neighbours eqn (4.63) takes the simple form

$$\frac{1}{T_1} = \frac{1}{T_1(\infty)} \frac{H_0^2 + 2H_{\mathrm{L}}^2}{H_0^2 + H_{\mathrm{L}}^2}.$$

5.2. *The spin temperature in the rotating frame* (Ref. 1, chap. 12; Refs. 30, 31)

The spin temperature concept was generalized in another direction by A. Redfield.[31] Let us consider the nuclear spin system in a high field H_0 submitted to a large oscillating field of amplitude $2H_1$ directed along the x-axis. It is convenient to describe the system in a frame of reference having the same z-axis but rotating around it with an angular velocity ω equal to the angular frequency of the oscillating field.

In this new frame the Hamiltonian becomes

$$\mathscr{H}' = \tilde{\mathscr{H}} + \text{terms varying in time,}$$

$$\frac{\tilde{\mathscr{H}}}{\hbar} = (\omega_0 - \omega) \sum_i I_{iz} + \gamma H_1 \sum_i I_{ix} + \frac{\mathscr{H}'_1}{\hbar}, \qquad (4.64)$$

where $\mathscr{H}'_1$ is the part of $\mathscr{H}_1$ that commutes with the operator I_z. Redfield made the assumption that a density matrix of the form (4.53), using $\tilde{\mathscr{H}}$ instead of $\mathscr{H}$, can still be used for describing the nuclear spin system, i.e.

$$\rho \simeq 1 - \frac{\tilde{\mathscr{H}}}{k_B \tilde{T}_S}, \qquad (4.65)$$

where $\tilde{T}_S$ is the spin temperature in the rotating frame.

We note that although the spin system is in a large static magnetic field, in the rotating frame the effective field H_{eff}, which is given by the relation $\gamma^2 H_{\text{eff}}^2 = (\omega_0 - \omega)^2 + \gamma^2 H_1^2$, may be small compared to the spin–spin interaction if ω is in the vicinity of ω_0.

A relaxation time $\tilde{T}_1$ is defined in the rotating frame as the time constant for the evolution of the inverse spin temperature; its value is given by an equation

$$\frac{1}{\tilde{T}_1} = \frac{1}{T_1(\infty)} \frac{\Delta^2 + H_1^2 + \tilde{\delta}\tilde{H}_L^2}{\Delta^2 + H_1^2 + \tilde{H}_L^2}, \qquad (4.66)$$

where $\gamma\Delta = \omega_0 - \omega$, thus $\Delta^2 + H_1^2 = H_{\text{eff}}^2$. $\tilde{H}_L$ is the local field calculated using the truncated spin–spin Hamiltonian $\mathscr{H}'_1$

$$\hbar^2 \gamma^2 \tilde{\mathscr{H}}_L^2 = \text{Tr}(\mathscr{H}'_1)^2,$$

and δ is defined as before to be the ratio of the relaxation rates for the spin–spin interactions $\mathscr{H}'_1$ and the Zeeman energy. More precisely, $\tilde{\delta}$ is defined by eqn (4.59) replacing $\mathscr{H}_1$ by $\mathscr{H}'_1$. The quantity $\tilde{\delta}$ has no reason to be equal to δ because the truncation changes the relative importance of the dipolar and scalar interactions. The exchange interaction is not affected by the truncation whereas the dipolar interaction is reduced. If the correlations are neglected $\tilde{\delta} = \delta = 2$.

The new local field has the following value

$$\hbar^2\gamma^2\tilde{H}_\mathrm{L}^2 = \frac{I(I+1)}{9}\sum_j B_{ij}^2 + I(I+1)\sum_j A_{ij}^2. \tag{4.67}$$

To establish eqn (4.66) we assumed that the Zeeman relaxation rates for the longitudinal term $\gamma\Delta I_z$ and for the transverse term $\omega_1 I_x$ are the same. This assumption is valid in metals because the Zeeman energies do not enter in the calculation of the relaxation rate.[32]

The value of $\tilde{\delta}$ may be expressed as a function of the ϵ_{ij} defined previously. The final result for $\tilde{T}_1$ is

$$\frac{1}{\tilde{T}_1} = \frac{1}{T_1(\infty)}\frac{\Delta^2 + H_1^2 + 2\tilde{H}_\mathrm{L}^2 + \sum_{i<j}\epsilon_{ij}(\tilde{H}_{\mathrm{d}ij}^2 - 2H_{\mathrm{s}ij}^2)}{\Delta^2 + H_1^2 + \tilde{H}_\mathrm{L}^2}, \tag{4.68}$$

with $\qquad\qquad \hbar^2\gamma^2\tilde{H}_{\mathrm{d}ij}^2 = \frac{1}{9}I(I+1)B_{ij}^2.$

Equation (4.68) is very useful because the experiments are easier in the rotating frame where the sample stays in a high field and the nuclear magnetization remains large. By studying the nuclear resonance line shape in the presence of a large radiofrequency field H_1,[33] or by measuring directly the rate of change of the spin temperature $\tilde{T}_\mathrm{S}$, it is possible to measure the quantities $\tilde{H}_\mathrm{L}$ and $\tilde{\delta}\tilde{H}_\mathrm{L}^2$.

5.3. The values of H_L, $\tilde{H}_\mathrm{L}$, δH_L^2, and $\tilde{\delta}\tilde{H}_\mathrm{L}^2$ in metals

5.3.1. The local field measurements. If the spin–spin interactions are purely dipolar, we may use eqns (4.44), (4.60), and (4.67) to relate the values of H_L and $\tilde{H}_\mathrm{L}$ to the second moment of the absorption line:

$$H_\mathrm{L}^2 = \tfrac{5}{3}(\Delta H)^2, \qquad \tilde{H}_\mathrm{L}^2 = \tfrac{1}{3}(\Delta H)^2,$$

where ΔH is the second moment expressed in magnetic field units. The measurement of H_L and $\tilde{H}_\mathrm{L}$ under these conditions does not provide new information, but the measurement can be used as a test of the spin temperature assumption.[32]

If a scalar interaction is present it contributes to the local fields but not to the second moment, therefore a measurement of H_L or $\tilde{H}_\mathrm{L}$ and of ΔH enables us to measure separately the two types of interaction. The measurement of $\tilde{H}_\mathrm{L}$ can be more useful because the dipolar contribution is reduced by the truncation and as this interaction is already known by the second moment measurement, the determination of a small scalar coupling will be more accurate if the value of $\tilde{H}_\mathrm{L}$ is used.

However, when the scalar interaction becomes much larger than the dipolar one, a difficulty arises.[33] In this case eqn (4.65) for the density matrix is not always valid. A more elaborate theory due to Provotorov,[34, 35] must be used. This theory, which will not be given here, predicts the nuclear line shape in conditions where the spin temperature assumption is not valid.[30]

In the case of two magnetic species the equations for the relaxation times for the two spins (in the laboratory or in the rotating frame) are

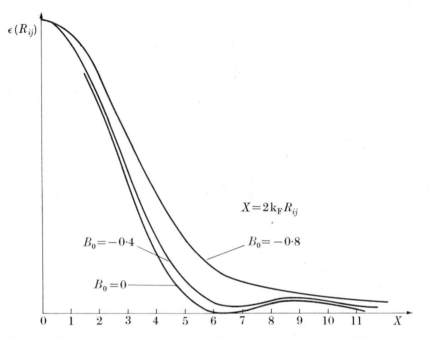

FIG. 4.4. Variation of the correlation coefficient $\epsilon(R)$ as a function of R, for several values of the electron–electron interaction coefficient B_0.

still valid provided a suitable redefinition of the quantities H_L, $\tilde{H}_L$, δH_L^2, and $\delta \tilde{H}_L^2$ is made. The calculations are straightforward but lengthy. The measurement of the local fields gives us the values of the scalar interaction with a better accuracy than a measurement of the second moment (a scalar coupling between unlike spins contributes to the second moment).

5.3.2. *Theoretical evaluation of the coefficient* ϵ_{ij}. This coefficient is given by eqn (4.61); its numerator has the value

$$\int_{-\infty}^{+\infty} d\tau \, \langle S_+(\mathbf{R}_i, t) S_-(\mathbf{R}_j, 0)\rangle,$$

which may be written after performing the integration over the time variable as

$$2\pi \sum_{i,j} P_i \delta(E_i - E_f)\langle E_i | S_+(\mathbf{R}_i) | E_f \rangle \langle E_f | S_-(\mathbf{R}_j) | E_i \rangle.$$

For non-interacting electrons, using the matrix elements of $S_-(\mathbf{R}_j)$ and $S_+(\mathbf{R}_i)$, the result

$$2\pi \sum_{\mathbf{k},\mathbf{k}'} \delta(E_{\mathbf{k}'} - E_{\mathbf{k}})|U_{\mathbf{k}}(\mathbf{R}_i)|^2 |U_{\mathbf{k}'}(\mathbf{R}_i)|^2 \exp\{i(\mathbf{k}-\mathbf{k}').(\mathbf{R}_i-\mathbf{R}_j)\} n_{\mathbf{k}}(1-n_{\mathbf{k}'})$$

is found. The Zeeman energy is neglected. The value of ϵ_{ij} is given by the equation

$$\epsilon_{ij} = \frac{\sum_{\mathbf{k},\mathbf{k}'} \delta(E_{\mathbf{k}'} - E_{\mathbf{k}})\delta(E_{\mathbf{k}'} - E_F)|U_{\mathbf{k}}(\mathbf{R}_i)|^2 |U_{\mathbf{k}'}(\mathbf{R}_i)|^2 \exp\{i(\mathbf{k}-\mathbf{k}').\mathbf{R}_{ij}\}}{\sum_{\mathbf{k},\mathbf{k}'} \delta(E_{\mathbf{k}'} - E_{\mathbf{k}})\delta(E_{\mathbf{k}'} - E_F)|U_{\mathbf{k}}(\mathbf{R}_i)|^2 |U_{\mathbf{k}'}(\mathbf{R}_i)|^2}.$$

$$(4.69)$$

If the quantity $|U_{\mathbf{k}}(\mathbf{R}_i)|^2$ is not too sensitive to the direction of the wave vector $\mathbf{k}$ at the Fermi surface, this quantity will be replaced by its average value at the Fermi surface and ϵ_{ij} will not depend upon the value of the wave function. Making this assumption, and assuming also a spherical Fermi surface, the summations over $\mathbf{k}$ and $\mathbf{k}'$ reduce to an integration over the directions of $\mathbf{k}$ and $\mathbf{k}'$ and the following result is obtained:

$$\epsilon_{ij} = \left(\frac{\sin k_F R_{ij}}{k_F R_{ij}}\right)^2. \qquad (4.70)$$

ϵ_{ij} is a rapidly decreasing function of R_{ij} presenting oscillations. As for the relaxation rate, ϵ_{ij} may be expressed using generalized susceptibilities. As ϵ_{ij} involves a correlation function for different spin positions, it will be related to $\chi_S(\mathbf{R}_i, \mathbf{R}_j)$. We get (see Appendix 1)

$$\epsilon_{ij} = \frac{\chi_S(\mathbf{R}_i, \mathbf{R}_j, \omega_n) - \chi_S(\mathbf{R}_j, \mathbf{R}_i, -\omega_n)}{\chi_S(\mathbf{R}_i, \mathbf{R}_i, \omega_n) - \chi_S(\mathbf{R}_i, \mathbf{R}_i, -\omega_n)}.$$

It is more convenient to use the Fourier transform of $\chi_S(\mathbf{R}_i, \mathbf{R}_i, \omega_n)$:

$$\epsilon_{ij} = \frac{\sum_{\mathbf{q}} \{\chi_S(\mathbf{q}, \omega_n) - \chi_S(-\mathbf{q}, -\omega_n)\}e^{i\mathbf{q}.\mathbf{R}_{ij}}}{\sum_{\mathbf{q}} \{\chi_S(\mathbf{q}, \omega_n) - \chi_S(-\mathbf{q}, -\omega_n)\}}.$$

If for $\chi_S(\mathbf{q})$ eqn (3.32) is used, ϵ_{ij} may be calculated as a function of R_{ij} by numerical integration.

The results of this calculation are shown on Fig. 4.4 for several negative values of B_0. The value of ϵ_{ij} is increased compared to the uncorrected value and the amplitude of the oscillations is reduced.

In Chapter 6 the few experimental results will be discussed and compared with the theory.

QUADRUPOLAR EFFECTS IN PERFECT METALS

1. Energy levels (Ref. 1, chap. 7)

IN Chapter 2 the expression for the Hamiltonian describing the interaction of the nuclear quadrupole moment with the electronic field gradients was established. The purpose of this section is to calculate the energy levels for a given spin, taking into account the Zeeman and quadrupolar interactions.

Using eqn (2.7), where a frame of reference diagonalizing the field gradient tensor was used, the total Hamiltonian is written as

$$\mathcal{H} = -\hbar\gamma_n \mathbf{H}_0 . \mathbf{I} + \frac{e^2 q Q}{4I(2I-1)} \{3I_z^2 - I(I+1) + \tfrac{1}{2}\eta(I_+^2 + I_-^2)\}. \quad (5.1)$$

In the most general case when the two interactions are comparable in strength it is necessary to solve a secular equation of order $2I+1$. For many experimental cases one of the interactions is large compared to the other and only such situations will be considered.

1.1. Weak quadrupolar coupling

The quadrupolar term is considered as a perturbation. It is now more convenient to take the z-axis along the direction of the magnetic field. For simplicity η will be assumed to be zero. The Hamiltonian becomes

$$\mathcal{H} = \hbar\omega_n I_z + \frac{e^2 q Q}{4I(2I-1)} [\tfrac{1}{2}(3\cos^2\theta - 1)\{3I_z^2 - I(I+1)\} +$$

$$+ \tfrac{3}{2}\sin\theta\cos\theta\{I_z(I_+ + I_-) + (I_+ + I_-)I_z\} + \tfrac{3}{4}\sin^2\theta(I_+^2 + I_-^2)], \quad (5.2)$$

where θ is the angle between the magnetic field and the old z-axis. A first-order perturbation calculation gives the following value for the energy levels:

$$E_m = \hbar\omega_n m + \frac{1}{8}\frac{e^2 q Q}{I(2I-1)}(3\cos^2\theta - 1)\{3m^2 - I(I+1)\}. \quad (5.3)$$

The most frequent situation met is when I is a half-integer. As the quadrupolar energy given by eqn (5.3) is only a function of m^2, the energy difference between the two levels $m = +\tfrac{1}{2}$ and $m = -\tfrac{1}{2}$ is not affected. This result has very important consequences. For a monocrystal several lines are observed and their separation is a measurement of the parameter e^2qQ. As in metals we are very often working with powders, an average over θ has to be taken. The resonance frequency of a given transition is given by the relation

$$2\pi\nu_{m\to m-1} = \omega_n + \frac{3e^2qQ}{8\hbar I(2I-1)}(2m-1)(3\cos^2\theta-1). \qquad (5.4)$$

The calculation of the line shape is completely similar to the calculation explained in the previous chapter for an anisotropic Knight shift. There

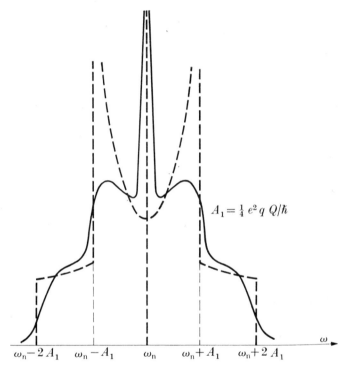

$$A_1 = \tfrac{1}{4}\,e^2q\,Q/\hbar$$

$\omega_n-2A_1 \qquad \omega_n-A_1 \qquad \omega_n \qquad \omega_n+A_1 \qquad \omega_n+2A_1$

FIG. 5.1. Line shape in a powder due to a first-order quadrupole splitting $I = \tfrac{3}{2}$. The full curve has another source of broadening superimposed.

is, however, a difference: several lines are present and the complete pattern is a superposition of the line shapes for the various possible values of m, the overall pattern is symmetrical around the component $m = \tfrac{1}{2}$, which is not affected by the first-order broadening. Such a pattern is presented in Fig. 5.1.

The broadening due to a first-order quadrupolar effect does not depend on the value of the external field. The observation of such a line shape provides a measurement of the quantity e^2qQ. If this quantity becomes too large only the central line is observed, the other components, usually called the satellite lines, being too broad to give a measurable signal. The central line itself may be affected by a broadening due to

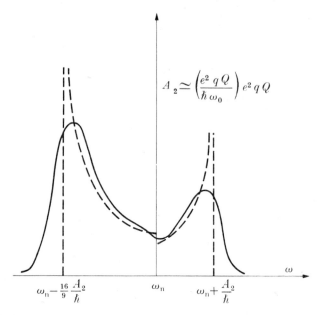

$$A_2 \simeq \left(\frac{e^2\,q\,Q}{\hbar\,\omega_0}\right) e^2\,q\,Q$$

$$\omega_n - \frac{16}{9}\frac{A_2}{\hbar} \qquad \omega_n \qquad \omega_n + \frac{A_2}{\hbar}$$

Fig. 5.2. Line shape due to a second-order quadrupolar broadening. In the full curve another source of broadening is taken into account.

second-order shift of the energy levels. The position of this line using a second-order perturbation calculation is given by the equation

$$h\nu_{\frac{1}{2}-\frac{1}{2}} = \hbar\omega_n + \frac{1}{\hbar\omega_n}\left(\frac{3e^2qQ}{2I(2I-1)}\right)^2\{I(I+1)-\tfrac{3}{4}\}(1-\cos^2\theta)(1-9\cos^2\theta).$$

$$(5.5)$$

In a powder such a shift will also produce a very asymmetrical broadening: the line shape is shown on Fig. 5.2. We note that the second-order broadening decreases if the magnetic field is increased. All these remarks about the field dependence of the line shape are useful to identify the origin of an observed line shape.

1.2. *Pure quadrupole resonance*

An opposite situation is met when the external magnetic field is reduced to zero.

The energy levels are only determined by the quadrupolar interaction and it becomes possible to observe sharp transitions without applying an external magnetic field. Such an experiment is called 'pure quadrupole resonance'. This technique is specially interesting in metals because the observed spectrum is the same in a powder as in a monocrystal; the measurement of the pure quadrupole resonance frequencies is a direct determination of the quadrupolar interaction.

For a site with axial symmetry the energy levels are

$$E_m = \frac{e^2qQ}{4I(2I-1)} \{3m^2 - I(I+1)\}. \tag{5.6}$$

The two levels with $I_z = \pm m$ are degenerate, so that if I has the value $\frac{3}{2}$ only one resonance is observed. As a matter of fact this degeneracy remains even if the symmetry is less than axial (but now because in the Hamiltonian the operators I_+^2 and I_-^2 are present the wave functions are no longer eigenfunctions of I_z).

For instance, when $I = \frac{3}{2}$ there is only one frequency given by the equation

$$h\nu = \frac{e^2qQ}{2}\left(1 + \frac{\eta^2}{3}\right)^{\frac{1}{2}}. \tag{5.7}$$

We notice that it is not possible using this technique to measure separately the two parameters e^2qQ and η. For spins larger than $I = \frac{3}{2}$ several resonance frequencies occur and the measurement of the two parameters becomes possible.

All these considerations are not restricted to metals; but in the calculation of the two parameters e^2qQ and η the contribution of the conduction electrons must be taken into account.

2. Quadrupolar relaxation rates

In this section we shall discuss the relaxation rates coming from the quadrupolar part of the hyperfine interaction. In Chapter 2 it was established that in a metal two contributions to the electric gradient are found, one is due to ionic charges, the other to conduction electron charges.

Both these contributions can produce a nuclear relaxation. The two terms connect the nuclear spins to different degrees of freedom; the ionic term couples the nuclear spins to the lattice vibrations, whereas the other term couples the electronic excitations to the nuclear spins.

The relaxation rate due to the coupling between the nuclear spins and the lattice vibrations (phonons) is, of course, not limited to metals, and will be mentioned here only for the sake of completeness.

2.1. *Phonon quadrupolar relaxation*

This process is very effective in ionic crystals and is characterized by a more complex temperature dependence than the electronic relaxation rates. This process is found to be negligible in metals.

2.2. *Electronic quadrupole relaxation rate*

For calculating this rate we consider only the part of the quadrupolar interaction coming from the conduction electrons.

The calculation is more conveniently done if the quadrupolar coupling (2.6) is expressed in terms of second-order tensors of the nuclear spin components. These operators are given by the equations

$$\left. \begin{aligned} Q_0 &= \frac{eQ}{2I(2I-1)}\{3I_z^2 - I(I+1)\} \\ Q_{\pm 1} &= \frac{eQ}{2I(2I-1)}\sqrt{\left(\frac{3}{2}\right)}(I_\pm I_z + I_z I_\pm) \\ Q_{\pm 2} &= \frac{eQ}{2I(2I-1)}\sqrt{\left(\frac{3}{2}\right)}I_\pm^2 \end{aligned} \right\}. \tag{5.8}$$

Using these quantities the quadrupolar coupling is written as

$$\mathcal{H}_Q = \sum_m Q_m V_{-m}, \tag{5.9}$$

V_m having the following values:

$$V_0 = \frac{1}{2}\frac{\partial^2 V}{\partial z^2}, \qquad V_{\pm 1} = \frac{1}{\sqrt{6}}\left\{\frac{\partial^2 V}{\partial x \partial z} \pm i\frac{\partial^2 V}{\partial z \partial y}\right\},$$

$$V_{\pm 2} = \frac{1}{2\sqrt{6}}\left\{\frac{\partial^2 V}{\partial x^2} - \frac{\partial^2 V}{\partial y^2} \pm 2i\frac{\partial^2 V}{\partial x \partial y}\right\}. \tag{5.10}$$

V is the electric potential created by the electrons.

The justification of the use of the form (5.9) is double. First, the matrix elements of the operator Q_m between nuclear Zeeman sub-levels are simple. Obviously Q_0 is diagonal, Q_{+1} is non-diagonal and connects only the states with $\Delta m = 1$, while Q_2 connects the states with $\Delta m = 2$. Secondly, if the quantities V_m are calculated as a function of the spherical coordinates r_e, θ_e, ϕ_e of the electron, it is found that they are simply expressed as functions of the second-order spherical harmonics:

$$V_m = -e\sqrt{\left(\frac{4\pi}{5}\right)}\sum_e Y_2^m(\theta_e, \phi_e)\frac{1}{r_e^3}. \tag{5.11}$$

These two properties are a consequence of the fact that the total Hamiltonian must be invariant if the axes of coordinates are rotated.

The non-diagonal terms of the quadrupolar coupling are able to produce a nuclear relaxation. Let us assume that the symmetry at the nuclear site is such that the static quadrupolar interaction vanishes. The calculation of the relaxation rate will be done as in the previous chapter by calculating the probability of transition for the electronic system due to the operators V_m (with $m \neq 0$). In fact the operators V_m are just those operators that appear in the calculation of the dipolar relaxation rate. As an example, the matrix elements of V_0 are exactly the quantities $\mathscr{T}^{zz}_{kk'}$ defined by eqn (4.11). The only difference is that the electronic spin variables are not involved in this calculation. The rate is found as for the other electronic rates to be proportional to $k_B T$. The calculations are performed in detail in an article by Mitchell. He assumes a spherical Fermi surface and expands the wave function taking only the s and p parts.[36]

The question remains to compare the magnitude of this relaxation rate with the dipolar relaxation rate. Neglecting the spin factors we have to compare e^2Q with $\hbar\gamma_n g\beta$. The ratio of the relaxation rates varies as the square of this ratio ρ.

Let us write $\hbar\gamma_n = ag\beta$, where a is a numerical constant equal to the ratio of the nuclear resonance frequency to the free-electron resonance frequency $(a \sim m/M$; m is the mass of the electron and M a nuclear mass).

$$\rho = \frac{e^2Q}{a(g\beta)^2} = \frac{m^2c^2}{a\hbar^2}Q = \frac{Q}{Q'},$$

with

$$Q' = a\left(\frac{\hbar}{mc}\right)^2 = 1\cdot2\times10^{-24}(a\,10^3)\ \text{cm}^2.$$

The quantities a and Q are given in various tables.

Let us discuss an example. For lithium $a = 0\cdot6\times10^{-3}$ and thus $Q' = 0\cdot7\times10^{-24}$ cm². For Q in cm² units, one finds $Q = 4\cdot2\times10^{-26}$ cm² and ρ is small and the quadrupolar rate is negligible. But in other cases ρ may be larger than one. The most striking example is found for ^{181}Ta where $Q' = 0\cdot35\times10^{-24}$ cm², whereas $Q = 2\cdot1\times10^{-24}$. Here the quadrupolar rate will be much larger than the dipolar rate.

Experimentally speaking, it is rather difficult to separate the various contributions to the relaxation rate. In principle, the magnitude of the Overhauser effect provides some information. As the quadrupolar rate does not involve the spin variables it does not produce an Overhauser enhancement. In the calculation of the coefficient ξ (eqn (4.31)) the quadrupolar rate is added to the total nuclear relaxation rate. Unfortunately this method can be performed only in very few metals having a very narrow electronic resonance line. Even for these metals the quadrupolar rate cannot be distinguished from the orbital relaxation rate, which also does not involve the spin variables.

Another method can be used when an element has several isotopes, with one or both having a quadrupole moment. In such a case the ratio of the magnetic parts of the total relaxation rate is known and is given by the equation

$$\frac{1}{T_{1a}}\bigg)_{\text{mag}} \bigg/ \frac{1}{T_{1b}}\bigg)_{\text{mag}} = \frac{\gamma_a^2}{\gamma_b^2}.$$

If the relation is not in agreement with the experimental results, this fact will prove that a quadrupolar relaxation rate is present. It is possible to determine in this case the ratio of the electric to the magnetic part of the interaction because the ratio of the quadrupolar part obeys the relation

$$\frac{1}{T_{1a}}\bigg)_{\text{quad}} \bigg/ \frac{1}{T_{1b}}\bigg)_{\text{quad}} = \frac{Q_a^2}{Q_b^2}.$$

3. Line shape in the presence of quadrupolar interactions

In this chapter the discussion is presented only for perfect metals and before considering the problem of the line shape it is fair to say that the discussion is rather academic, since the source of the line width when $I > \frac{1}{2}$ is very often the quadrupolar effects around defects or impurities. This important problem will be discussed in Chapter 7. The question that will be discussed here is the modification of the spin–spin broadening produced by the presence of a static quadrupolar interaction. If the symmetry at the nuclear site is not cubic and if a powdered sample is used, the quadrupolar splitting usually produces for all the satellite lines a broadening much larger than the spin–spin broadening. The only question to be solved is what happens to the width of the central line due to the spin–spin interactions. We assume that the second-order quadrupole broadening is negligible.

The method of moments can still be used provided the following modifications are made.

(1) First a new truncated spin–spin Hamiltonian has to be defined. In the absence of quadrupolar terms the truncated Hamiltonian was the part of the spin–spin interactions $\mathscr{H}_I$ that commutes with the total Zeeman energy; now we must use the part of $\mathscr{H}$ that commutes with the Zeeman and quadrupolar energies. If the Zeeman energy is large the Hamiltonian given by eqn (5.2), neglecting the non-diagonal terms, may be written as

$$\mathscr{H} = \hbar \omega_n I_z + \delta \sum_i (I_{iz})^2. \tag{5.12}$$

As δ is a function of the angle between the field and the crystalline axis this Hamiltonian is valid for a given crystallite. In a powdered sample δ varies from one crystallite to the other.

We assume that δ is larger than the spin–spin interactions for almost all crystallites. In $\mathscr{H}_I$ the only terms commuting with the Hamiltonian given by eqn (5.12) are terms like $I_{jz} I_{iz}$ and among the terms $I_{i+} I_{j-}$ the terms coupling a state $m_i = m$, $m_j = m \pm 1$, to the state $m_i = m \pm 1$, $m_j = m$ (because for these two states the value of $m_i^2 + m_j^2$ is not changed).

(2) The operator I_x used in eqns (4.42) and (4.43) is also changed; only the part of I_x connecting the states $m = \pm \frac{1}{2}$ has to be considered.

The result of the moment calculation is that a small change is obtained for the value of M_2.[37]

The same kind of discussion applies to other forms of the Hamiltonian, including the pure quadrupole resonance situation. This discussion is not at all dependent on the fact we are dealing with a metal.

4. Ultrasonic excitation[38]

In Chapter 1 we said that in a nuclear resonance experiment the signal is observed by measuring a change in the properties of an electromagnetic perturbation. The transitions between the nuclear sublevels are induced by a time-varying magnetic field. When the nucleus possesses a quadrupole moment there is another possible way of performing a nuclear resonance experiment. Let us assume that in a metal (or more generally in a solid) an ultrasonic wave is sent, with a frequency in the vicinity of a nuclear resonance frequency. The ultrasonic wave induces a motion of the ions, producing, therefore, a coherent modulation of the quadrupolar Hamiltonian. More precisely, the quantities V_m defined by eqn (5.10) have a part that is time oscillating and they will behave as an oscillating magnetic field. For observing this effect two types of experiments are possible.

(1) It is possible in principle to measure the ultrasonic attenuation as a function of the frequency (or of the magnetic field). Such an experiment is completely similar to the usual nuclear magnetic resonance method. The selection rules for the ultrasonic transitions are different because the operator $V_2 Q_2$ in the Hamiltonian induces a transition with $\Delta m = 2$ (transitions with $\Delta m = 1$ are also induced, using the operator $V_1 Q_1$). This remark is important for experimental reasons, because a transition $\Delta m = 2$ cannot be induced by the electromagnetic source used to produce the ultrasonic wave.

(2) Another possibility is as follows. A very powerful ultrasonic wave is sent and this wave equalizes the populations of the nuclear sub-levels. The change of populations is detected by a change in the intensity of the signal observed with a radiofrequency field.

The possibility of ultrasonic experiments is specially interesting in a metal because the ultrasonic wave does not suffer the skin depth attenuation. Nuclear acoustic resonances were observed in some metals, the most interesting case being the observation of the signal in metallic ^{181}Ta; we have already noted that the quadrupole moment of this nucleus is rather large.[39]

Finally, let us mention that recently a nuclear acoustic resonance was observed which does not involve quadrupolar coupling.[40] When an ultrasonic wave is sent in a metal this wave induces a motion of the conduction electrons; in the presence of a magnetic field the motion produces an electromagnetic wave oscillating at the ultrasonic frequency. A nuclear magnetic resonance due to this wave may be observed. This resonance involves the coupling with a nuclear magnetic moment and obeys the usual selection rule $\Delta m = 1$. Such an effect was observed in aluminium.

This mechanism is interesting because, as the wave is created inside the metal, the skin effect does not appear. It becomes possible to study the resonance in a large metallic single crystal for nuclear spins $I = \frac{1}{2}$ or having very weak quadrupole moment.

6

RESULTS OF NUCLEAR MAGNETIC
RESONANCE EXPERIMENTS IN METALS

In this chapter we shall review some of the experimental results and compare them with the theoretical predictions established in Chapters 4 and 5. The discussion will be limited to normal metals with no incomplete internal shell. We will not try to give a complete survey of all the results, but restrict the discussion to the cases where a detailed comparison with the theory is possible.

1. The alkali metals

These metals are the simplest example; they have only one very wide half-filled energy band and a model starting from plane wave functions is well fitted to describe conduction electron properties. They crystallize in a cubic system so that none of the inhomogeneous broadening due to anisotropic Knight shift or quadrupolar splitting are present. The Fermi surface of the alkali metals is not too different from a sphere, especially for sodium, potassium, and rubidium. (This statement is less correct for lithium and caesium.)

The electronic resonance of conduction electrons has been observed for all the alkali metals and is found to be very narrow in sodium and lithium.

The indirect interactions are negligible for lithium and sodium but they become large for the two heavier metals, rubidium and caesium.

The nuclear resonance of potassium has received much less attention, since the low value of the nuclear moment reduces the signal-to-noise ratio compared to what is observed in the other alkali metals.

1.1. *The measurement of* χ_S *and* $|U_{k_F}(\mathbf{R}_i)|^2$

If we assume at first that the Knight shift has no noticeable orbital contribution (there is no possibility of a dipolar contribution as a result of the cubic symmetry), its measurement gives us the product of the two quantities χ_S and $|U_{k_F}(\mathbf{R}_i)|^2 = |\phi_{k_F}(\mathbf{R}_i)|^2$. For lithium and sodium it is possible to measure χ_S separately and so the two quantities are known. χ_S is determined by measuring the intensity of the conduction electron resonance line. We noticed in Chapter 4 that eqn (4.3) is also valid for the electronic spin system, and the following equation is obtained:

$$\chi_S = \frac{1}{\pi} \mathscr{P} \int\limits_{-\infty}^{+\infty} \frac{\chi_S''(\omega)}{\omega} \, d\omega. \tag{6.1}$$

An absolute measurement of χ_S'' would not be a simple experiment as it would require a knowledge of the absolute value of the exciting variable field and of the absolute value of the variable electronic magnetization M_x. To avoid these difficulties, the nuclear resonance signal $\chi_n''(\omega)$ is measured with the same variable field H_1 (the nuclear signal for ^{23}Na or ^{7}Li). The two functions $\chi_S''(\omega)$ and $\chi_n''(\omega)$ are integrated and the ratio of these two results χ_S/χ_n is taken. As χ_n is known, χ_S is measured. Because the same frequency is used for the electronic and nuclear experiments, the electronic signal is observed in a very small static magnetic field, and the experiment is only possible for a narrow electronic line (narrower than the applied field).[41, 42] The experimental values for χ_S are found to be larger than the values for a free-electron gas given by eqn (3.19). This result is expected, for a repulsive Coulomb interaction gives a negative value for the coefficient B_0. If we assume that there is no orbital contribution to the Knight shift, the values of $|\phi_{k_F}(\mathbf{R}_i)|^2$ are deduced. The experimental values for χ_S and $|\phi_{k_F}(\mathbf{R}_i)|^2$ are given in Table 6.1 and compared with several theoretical predictions.

For sodium the shift of the electronic resonance line can be measured at low temperatures. We measure the change of the position of the line when the nuclear magnetization is reduced to zero by applying a large radiofrequency field. Using eqn (4.8) a direct determination of $|\phi_{k_F}(\mathbf{R}_i)|^2$ is obtained.

The two determinations are in good agreement with each other. This result is a proof of the lack of orbital contribution for this metal.[45]

1.2. *Variation of the Knight shift with pressure and temperature*

Another problem is to study the variation of the Knight shift when external parameters such as the pressure or the temperature are changed. Let us first consider the influence of the temperature. It is important to keep in mind that in all non-transition metals, the variations of the Knight shift with the temperature are usually small. As an example, for caesium, where the temperature dependence is considered as large, the Knight shift varies by about 6 per cent when the temperature is rising from 1 to 300 K.

When the temperature changes, because there is a thermal expansion of the solid, the size of the unit cell changes and this variation modifies the values of the two quantities χ_S and $|\phi_{k_F}(\mathbf{R}_i)|^2$. But the influence of the change in the size of the unit cell can be measured by looking at the variation of the Knight shift with pressure. The variation as a function

TABLE 6.1

Results of the experiments for χ_S and $|\phi_{k_F}(\mathbf{R}_i)|^2$ and the theoretical predictions

	Ref.	Li	*Ref.*	Na	*Ref.*
$10^6\chi_S$ experiments		$2\cdot08\pm0\cdot1$	41	$0\cdot95\pm0\cdot1$	41
		$1\cdot96\pm0\cdot1$	43	$1\cdot13\pm0\cdot1$	42
$10^6\chi_S$ theory Pauli		$1\cdot17$		$0\cdot65$	
	15	$5\cdot3$		$1\cdot7$	
	14	$1\cdot87$		$0\cdot85$	
	44	$2\cdot12$		$0\cdot97$	
$|\phi_{k_F}(\mathbf{R}_i)|^2/|\phi_A(\mathbf{R}_i)|^2$ experiments		$0\cdot46$ (a)		$0\cdot63$	45
				$0\cdot60$ (a)	
$|\phi_{k_F}(\mathbf{R}_i)|^2/|\phi_A(\mathbf{R}_i)|^2$ theory	46	$0\cdot49$		$0\cdot80$	
	47	$0\cdot455$		$0\cdot79$	

$|\phi_A(\mathbf{R}_i)|^2$ is the electronic density at the nucleus for the free atom; the susceptibility χ_S is expressed in C.G.S. per unit volume.

(a) This value is deduced from the Knight shift and χ_S assuming no orbital contribution to the Knight shift.

of the pressure of the Knight shift for the alkali metals was measured by Benedek.[48] In analysing the results we are faced with the difficulty found when the absolute value of the Knight shift was discussed, namely, only the variation of the product $\chi_S|\phi_{k_F}(\mathbf{R}_i)|^2$ is measured and there is no experimental way to separate the two variations. The results were analysed by using a theoretical prediction for the variation of χ_S with the volume of the cell. Benedek uses for the variation of χ_S a theory due to Pines,[14] which takes into account the interactions between electrons. The choice of this theory was justified by the fact that this model explains rather well the absolute measured value of the susceptibilities for Na and Li (see Table 6.1). We are able therefore to deduce the variation of $|\phi_{k_F}|^2$ with the volume, knowing the compressibility with the pressure. The results are compared with the theoretical predictions using wave functions calculated by Brooks. The agreement is rather good for sodium, lithium, and rubidium but is less satisfactory for caesium. From the experimental point of view it is important to realize that these experiments require a very accurate determination of the position of the resonance; we are looking at a small variation of a small shift. The measurements are possible only because the line width at room temperature is considerably reduced by ionic diffusion.

It is now possible to compare the experimental thermal variation of the Knight shift with the expected variation due to thermal expansion and it is found that the temperature dependence is not entirely accounted for by the thermal expansion effect. There is a temperature dependence at constant volume. The physical explanation of this effect is as follows. In an experiment performed at a temperature T, there are lattice vibrations that give rise to oscillating time variation of the volume of the cell, say $\Delta v(t)$. Let us expand the electronic density $|\phi_{k_F}(\mathbf{R}_i)|^2$, which we shall call P_F, as a function of Δv. We find

$$P_F\{V_0+\Delta v(t)\} = P_F(V_0)+\left|\frac{\partial P_F}{\partial V}\right|_{V_0}\Delta v(t)+\tfrac{1}{2}\left|\frac{\partial^2 P_F}{\partial V^2}\right|_{V_0}\{\Delta v(t)\}^2, \qquad (6.2)$$

where V_0 is the average volume. If the time average of this quantity is taken, and as $\overline{\Delta v(t)} = 0$ (otherwise the average volume will differ from V_0), the following result is obtained:

$$P_F(V_0, T) = P_F(V_0)+\tfrac{1}{2}\left|\frac{\partial^2 P_F}{\partial V^2}\right|_{V_0}\overline{\Delta v_T^2}, \qquad (6.3)$$

where $P_F(V_0, T)$ is the measured value of the electronic density, $P_F(V_0)$ is the density in the absence of thermal vibrations, and $\overline{\Delta v_T^2}$ is the thermal average of $\{\Delta v(t)\}^2$. $P_F(V_0)$ is not the value of $P_F(V_0, T)$ for $T = 0$ K because even at $T = 0$ K there exist zero-point vibrations and $P_F(V_0, 0)$ is given by the relation

$$P_F(V_0, 0) = P_F(V_0)+\tfrac{1}{2}\left|\frac{\partial^2 P_F}{\partial V^2}\right|_{V_0}\overline{\Delta v_0^2}, \qquad (6.4)$$

or $$P_F(V_0, T) = P_F(V_0, 0)+\tfrac{1}{2}\frac{\partial^2 P_F}{\partial V^2}\{\overline{\Delta v_T^2}-\overline{\Delta v_0^2}\}.$$

The presence of the zero-point vibrations is experimentally proved by the following experimental fact. If the Knight shifts of the two isotopes of lithium, ^{6}Li and ^{7}Li, are compared (at very low temperature) they are found to be slightly different.[49] As the electronic susceptibilities are the same, we have to admit that a change in the electronic density $P_F(V_0, 0)$ appears. This change arises because the zero-point motion is different for the two isotopes due to the relatively large change in the ionic masses. In lithium the change in $(\overline{\Delta v^2})_0$ is quite noticeable.

1.3. Overhauser effect and relaxation time

There are many measurements of the nuclear relaxation rate in alkali metals. This rate is well explained as a superposition of two mechanisms:

a modulation of the magnetic hyperfine interaction by the electronic motion, and a modulation of the spin–spin coupling by the diffusion process. This second contribution is important only at high temperature. The Overhauser effect was observed and measured for lithium and sodium, and this question will be discussed now in more detail. The coefficient ξ defined in Chapter 4 (eqn (4.31)) is known in both these

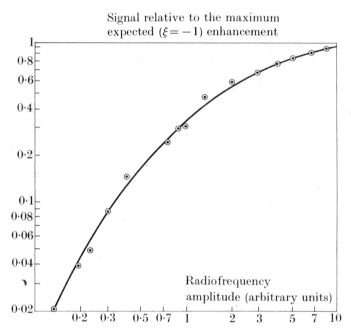

Signal relative to the maximum expected ($\xi = -1$) enhancement

Radiofrequency amplitude (arbitrary units)

FIG. 6. 1. Variation of the nuclear magnetization as a function of the intensity of the saturating field (for ^{23}Na).

metals. The experiment was performed in a low magnetic field,[50] since to observe a complete Overhauser effect ($s = 1$) it is extremely important to have a sample whose particles are smaller than the skin depth. This condition is easily satisfied by using a low-frequency saturating field.

In sodium it was found that $\xi = -1$, which means that the relaxation rate arises entirely from the scalar interaction. In Fig. 6.1 the amplitude of the nuclear resonance signal is plotted as a function of the amplitude of the saturating field at the electronic resonance frequency. As we explained earlier in this chapter, the Knight shift too is due to the hyperfine contact interaction, thus we expect a rather good verification of the corrected form (4.28) of the Korringa relation. However, this corrected form neglects the effect of the electron–electron interactions on the relaxation rate. The deviation from eqn (4.28) may be used to estimate the effect of the interactions on the relaxation.[51] It is found that the corrections are smaller than those given by the theory of Wolff (eqn (3.32)), which assumes an electron–electron interaction of very short range. Comparable results are found in lithium. For this metal the measurement of the Overhauser effect leads to the conclusion that the relaxation rate at low temperature is not entirely due to the scalar term. The value of ξ was found to be $\xi = 0 \cdot 93 \pm 0 \cdot 03$.[50] To explain the results, the authors assumed a spherical Fermi surface and a wave function with only p and s components. With these assumptions the dipolar relaxation rate is found to be more than an order of magnitude smaller than the orbital relaxation rate (the quadrupolar relaxation rate is also negligible). Using wave functions calculated by Callaway and Kohn[52] the orbital relaxation rate is found to be 6 per cent of the scalar rate, in very good agreement with the measured value of ξ. For these two metals the shift of the electronic line due to the enhanced nuclear magnetization has been observed. These shifts are very large, for example, if $\langle I_z \rangle = I$, i.e. complete nuclear polarization. (Such a situation would be realized by a complete Overhauser effect in a large field $H_0 \sim 25\,000$ G and at low temperature $T \sim 1$ K.) Equation (4.8) gives the following values:

$$^7\text{Li} \quad H_\text{n} = 80 \text{ G}, \qquad ^{23}\text{Na} \quad H_\text{n} = 310 \text{ G}.$$

The experimental results are in good agreement with the theoretical predictions[53] (and Ref. 1, chap. 9).

1.4. *Indirect interactions and the correlation coefficient* δ

Although no indirect spin–spin interaction is observed for the two lighter metals, it is comparable to the classical dipolar coupling for rubidium and very large for caesium.

Thus the values of H_L or $\tilde{H}_\text{L}$ are well known for sodium and lithium, and if an accurate measurement of the relaxation rate as a function of the magnetic field (in the laboratory or in the rotating frame) is performed, the coefficient δ can be measured.

1.4.1. *Measurement of* δ *in sodium and lithium.* The first experiment was performed by Redfield[54] by measuring the variation of T_1 as a function of the external field. The aim of this experiment was to test the validity of the spin temperature assumption and not to measure δ with great accuracy. As a matter of fact the important quantity to compare with the theoretical predictions is not δ but ϵ_{ij} (it will be very often

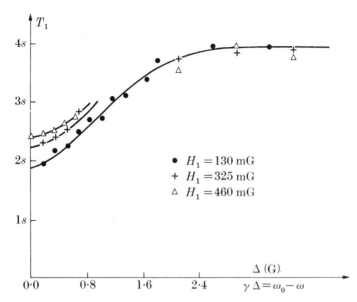

FIG. 6.2. This figure shows the variation of T_1 as a function of Δ for several values of the amplitude H_1 of the rotating field. The value of δ is obtained by extrapolating T_1 ($\Delta = 0$) for a zero value of H_1.

TABLE 6.2

Theoretical and experimental values of δ for sodium and lithium

	Li	Ref.	Na	Ref.
δ experiments	2·20	54	2·28	54
			2·15±0·06	55
	2·31±0·05	56	2·13±0·03	56
δ theory				
free electrons			2·011	2·011
Wolff $B_0 = -0·4$	2·028		2·028	

assumed that ϵ_{ij} is only important for nearest neighbours and when we write ϵ_{ij} we mean the value of ϵ_{ij} for nearest neighbours). This quantity is a function of $\delta-2$ and to obtain a value of ϵ_{ij}, δ must be known with great accuracy.

More recent experiments were performed either in the laboratory frame but using a different experimental technique,[55] or in the rotating frame.[56] In Fig. 6.2 the variation of T_1 as a function of the field in the rotating frame is shown.

The results, given in Table 6.2, show that all the experimental values of δ are larger than the theoretical predictions (eqn (4.70) and Fig. 4.4).

The free-electron model with a spherical Fermi surface (4.70) leads to an extremely small value of ϵ_{ij}, because in a body-centred cubic lattice the distance R_{ij} between the first neighbours gives for the product $k_F R_{ij}$ a value nearly equal to π.

As mentioned in Chapter 4, the Wolff theory[23] predicts a larger value for ϵ_{ij}. Let us consider eqn (3.24), which gives the static susceptibility. Knowing χ_S, we may use this equation to estimate B_0. If we assume $m^* = m$ we find for both metals the value $B_0 = -0\cdot4$. (Since the ratio m^*/m is larger than one, this determination considerably overestimates $|B_0|$; in Chapter 10 it will be shown that for sodium a better choice would be $B_0 = -0\cdot2$.) With this value ($B_0 = -0\cdot4$) we obtain for δ the theoretical values presented in Table 6.2, which remain much smaller than the experimental results. This crude theory also predicts the same value of δ for sodium and lithium, whereas the experimental values are rather different. Several effects may explain this difference. First, the ratios m^*/m are different. In lithium this ratio deduced from specific heat measurements is about 2, so that a small but positive value of B_0 is expected. This explanation is not able to explain the result for lithium. The difference may also be due to the fact that the Fermi surface for lithium is not a sphere. Finally, there remains the possibility of the contribution to the local field of quadrupolar splitting due to imperfections. If there is such a contribution H_Q^2 to the local field, it can be shown that for a quadrupole interaction a value of $\delta = 3$ is obtained. This result may be explained as follows. A quadrupolar interaction $(I_{iz})^2 - \frac{1}{3}I(I+1)$ has exactly the form of a dipolar interaction $I_{iz}I_{jz} - \frac{1}{3}\mathbf{I}_i.\mathbf{I}_j$ with $i = j$. Consequently if the relaxation rate is calculated one finds for $\delta: \delta = 2 + \epsilon_{ii} = 3$. Thus the presence of quadrupolar interactions will always increase the value of δ.

1.4.2. *Measurement of indirect interactions in rubidium and caesium.* These two metals offer a very good example of an experimental determination of indirect interactions, using the results of the theory based on the spin temperature concept.

In rubidium two isotopes, ^{85}Rb and ^{87}Rb, are present in relative natural abundances of 72·8 and 27·2 per cent respectively. It is possible to change these figures by using samples enriched in one of the isotopes. For interpreting the results a value $\epsilon_{ij} = 0$ will be assumed. This simplifying choice may be justified by the following remarks. The result of the measurements on the lighter alkali metals shows that ϵ_{ij} is not very large. When two isotopes are present a large contribution to the local field comes from the interaction between unlike spins $\mathcal{H}_{IS}$,

$\mathscr{H}_{IS} = \sum_{i,j'} C_{ij'} I_{iz} S_{j'z}$. If the relaxation rate for this interaction is cal-culated using eqn (4.56), we find that because the trace is taken over the nuclear spin variables the cross terms involving the correlation function $S_+(\mathbf{R}_i, t)S_-(\mathbf{R}_{j'}, 0)$ disappear and the rate becomes simply $1/T_I + 1/T_S$, where T_I and T_S are the relaxation times for the Zeeman energies of the spins I and S. ϵ_{ij} does not appear in the relaxation rate of this interaction.

Finally, if we consider the relaxation rate for the interaction between like spins, it is found that in rubidium the magnitudes of the scalar and dipolar contributions are comparable. Looking at eqn (4.63) we see that ϵ_{ij} is multiplied by $H^2_{\mathrm{d}ij} - 2H^2_{\mathrm{s}ij}$, and the value of the total relaxation rate is very insensitive to the presence of an ϵ_{ij} having a small value.

By measuring the local fields for the two isotopes it is possible to measure the scalar and dipolar parts of the interactions in rubidium.[57,58] The results are given in Table 6.3. We shall discuss the theoretical predictions for the interactions. The simplest possibility is to use the Ruderman–Kittel equation (4.49). The interaction is a function of two parameters $|U_{k_F}(\mathbf{R}_i)|^2$ and the effective mass m^*. If for m^* the value deduced from the specific heat measurement is taken[59] $m^* = 1\cdot26m$, and if for the electronic density the value for atomic rubidium is assumed, a value in very good agreement with the experimental results is found. This agreement may be accidental, due to the large number of approxi-mations in the theory. There are other calculations, using more realistic wave functions,[60-2] which are also given in Table 6.3. In the table the value of the pseudo-dipolar coupling is given. This value is in serious disagreement with the calculations of References 61 and 62; the experi-mental value of B_{IS} is much too large. The accuracy of the measurement of this quantity is very poor. There is also the possibility of a contribution to the local field from quadrupolar splitting due to impurities, and such effect will considerably change the value of B_{IS}. In References 57 and 58 it was shown that the introduction of a pseudo-dipolar term was motivated by the large value found for the absorption line width in an enriched sample (97 per cent of ^{85}Rb) but such an interaction does not affect appreciably the value of the local fields. Any other source of line width will therefore change considerably the experimental value of B_{IS} but not of A_{IS}.

A different situation is found in caesium, where only one isotope is present. Due to the large value of the nuclear charge a strong indirect coupling is expected, leading to a very narrow line. (This result is valid

TABLE 6.3

The quantities A_{IS} and B_{IS} for Rb and Ag are the scalar and pseudo-dipolar interactions for unlike-spin nearest neighbours. A_{II}, B_{II} are the same quantities for like-spin nearest neighbours. All the figures are expressed in hertz.

	Rb				Cs				Ag	
Ref.	A_{IS}	Ref.	B_{IS}	Ref.	A_{II}	Ref.	B_{II}	Ref.	A_{IS}	Ref.
Results	174±14	57, 58	~40	57, 58	200±20	57, 58	~35	57, 58	26.5±1.5	63
Theory R.K. 28	187				324				40	
60	500				830					
61	72		1.35		124		2.3			
62	83		1.5		190		2.65			

only at low temperature; at high temperature in the vicinity of the melting point the relaxation rate is a source of broadening.) The observed line is indeed rather narrow. However, the experiments show that the line width is very sensitive to the purity of the sample and the various experimental determinations are not always in agreement. This behaviour is easy to understand. As will be shown in the next chapter, the presence of impurities in a metal changes the value of the Knight shift for the nuclear moments in the vicinity of an impurity. The Knight shift value in caesium is very large ($\sim 1\cdot 5$ per cent) and even a small amount of impurities produces a broadening. This broadening is easily identified because it is proportional to the external field, and the intrinsic line width can be measured by plotting the experimental width as a function of the field and extrapolating its value to the zero applied field.

As stated in Chapter 4, it is not possible to deduce the scalar and dipolar interactions by measuring only the width of the exchange narrowed line. On the other hand, a measurement of the local field $\tilde{H}_{\mathrm{L}}$ is also difficult when the exchange part of the interaction becomes large.[33,57,58] The Provotorov theory enables us to calculate in detail the shape of the dispersion signal. In this line shape the magnitudes of the dipolar and scalar couplings enter as parameters and thus may be deduced from the experimental line shape (all the details are found in Reference 56). The results are given in Table 6.3 and compared with the Ruderman–Kittel calculation. (Here again the effective mass $m^* = 1\cdot 26m$ is deduced from the specific heat and the electronic density is the density for atomic rubidium.) We note in Table 6.3 the rather good agreement found between the experimental value of A_{II} and the result of the calculations of Mahanti, Das, and Tserlikkis.[61,62] For caesium, too, a serious disagreement is found for the value of the pseudo-dipolar interaction. The explanation based on a possible quadrupole broadening is very doubtful because the quadrupole moment of caesium is very small.

Finally, let us mention that there are several measurements of the caesium resonance line width in the vicinity of the melting point but the interpretation of the results is rather complicated. There are two new effects, a diffusion process that tends to average the spin–spin interaction and a contribution of the relaxation rate to the line width.

2. The noble metals

There are some similarities between these three metals and the alkali metals: they are monovalent and they crystallize in a cubic system (but a different one, the face-centred cubic system). However, there are also serious differences: the Fermi surface is more complicated, its shape being very far from a sphere, and the quadrupolar effects are important in copper and gold. A resonance line due to the conduction electrons has been observed in silver and copper.

2.1. *Resonance in silver*

The measurement of the line width in silver was the first experiment proving the existence of indirect interactions. Silver has two isotopes, ^{107}Ag and ^{109}Ag, both with $I = \frac{1}{2}$. The situation presents some similarity to the one discussed for rubidium. The indirect interactions are measured with a good accuracy by measuring the local fields.[63] The results are given in Table 6.3 where it is seen that the indirect interaction is somewhat smaller than the Ruderman–Kittel value.

Many studies of nuclear resonance in alloys were made using silver as the solvent, this metal being rather easy to alloy and presenting no quadrupolar effects.

2.2. *Resonance in copper*

Here also two isotopes are present, ^{63}Cu and ^{65}Cu, both having a quadrupole moment. Very recently a measurement of the electronic spin susceptibility was reported[65, 66] (the method will be discussed in Chapter 10).

The variation of the relaxation rate as a function of the external magnetic field has also been measured. The variation does not follow the behaviour expected from eqn (4.57).[54] The authors found that the ratio of the relaxation rates in a zero field to the rate in a large field is surprisingly large, $\delta = 2 \cdot 72$, and also that the measured local field cannot be explained by dipolar interactions. These effects very likely arise from local quadrupolar splitting due to imperfections (as explained earlier a pure quadrupole interaction will give $\delta = 3$). As in the case of silver, many experiments in alloys are done using copper as the solvent. With copper the effects will be of a quadrupolar origin. This question will be discussed in the next chapter.

2.3. *Resonance in gold*

Resonance was observed using a very large magnetic field.[67] The nuclear moment of ^{197}Au is very small; in a field of 10 000 G the resonance frequency is 741 kHz. The line width, which is much larger than the dipolar line width, is very likely due to quadrupolar effects.

3. Other metals

We shall briefly discuss the more striking results obtained in the other metals without attempting a complete enumeration. In most of the cases only the Knight shift and the relaxation rate are known.

3.1. *Divalent metals*

These metals are characterized by a rather small value of the density of states at the Fermi energy. Therefore we expect relatively small values for the electronic susceptibility, the Knight shift, and the relaxation rate.

The Fermi surfaces have a complicated shape. The symmetry is not cubic. The first divalent metal is beryllium, which behaves in a very peculiar way. Its Knight shift was known to have a very small value, much smaller than expected from an estimate of the susceptibility. Recently[68] this shift was determined to be negative, a result sometimes found for transition metals. This fact suggests that there is a shift from a 'core polarization effect' which will be discussed in Chapter 8. A theoretical attempt[69] to calculate the Knight shift gives a value in disagreement with the experiment. The calculation is difficult due to the complicated shape of the Fermi surface. The authors suggest the possible occurrence of an important negative orbital contribution to the shift.

As already mentioned, the electronic susceptibility is known (though with poor accuracy) and is about an order of magnitude smaller than that for sodium.

Beryllium metal does not have a cubic structure and a static quadrupole splitting is observed.[70] Due to the small number of electrons in the inner shell the calculation of the electric field gradient in beryllium is not too complicated. The first calculation was in moderate agreement with experiment, the part of the gradient due to the conduction electrons being estimated to be about 8 per cent of the total gradient.[71]

This last contribution was recently re-estimated to be about -17 per cent of the total gradient, which leads to a better agreement with the experimental value.[72]

The static quadrupole splitting was also measured for magnesium and is in reasonable agreement with theory.[73] In cadmium a noticeable anisotropic Knight shift is observed.[74] For this metal both the isotropic and anisotropic shifts present a large temperature dependence[75-7]. The isotropic shift increases by about 70 per cent between 4 and 594 K (which is the melting temperature) (Fig. 6.3). The anisotropic shift is small and negative at 4 K; after going through zero it increases to a rather large positive value at the melting temperature. The Fermi surface of cadmium is rather complicated; as the temperature increases the lattice vibrations tend to decrease the value of the periodic lattice potential and the wave functions become more and more comparable to free-electron wave functions. The density of states (and therefore the susceptibility and the Knight shift) increases strongly when the temperature is increased. A detailed calculation is in rather good agreement with the experimental results.[78]

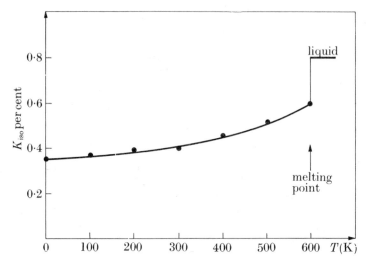

FIG. 6.3. Variation of the isotropic part of the Knight shift (K_{iso}) in ^{113}Cd as a function of temperature (Refs. 75–8).

3.2. Trivalent metals

The nuclear resonance properties of ^{27}Al present some similarity with the properties found for copper (except for the fact that only one isotope is present). The quadrupolar effects are important and, as in copper, the behaviour of the relaxation rate as a function of the field gives a large value for δ and also a large value for the local field.[54] Systematic measurements of δ as a function of the purity and of the temperature were performed more recently.[79] The results show the importance of quadrupolar effects.

Recently an electronic resonance line was observed in aluminium. Many studies on alloys are performed using aluminium alloys.

Nuclear resonance was studied in detail in thallium. This study is of historical interest because it was one of the first examples of an exchange narrowing. There are two isotopes with spins $I = \frac{1}{2}$. We observe the presence of large indirect interactions as for rubidium. However, the symmetry is not cubic and a part of the broadening comes from the anisotropy of the Knight shift.

In gallium and indium a pure quadrupole resonance spectrum is observed. In the case of gallium a careful study of the pressure and temperature dependence of the splitting was made.[80] Here again the largest contribution to the field gradient comes from the ionic charges.

3.3. *Metals of the fourth column*

In a study of lead the indirect interactions were measured using a technique that will be described in the next chapter.[72] The temperature dependence of the Knight shift in tin was also investigated. Here also an anisotropic shift is observed and the results are somewhat similar[81] to the result for cadmium. However, the thermal variations are much smaller and the change in the isotropic shift on going from 4 to 500 K is only 5 per cent.

NUCLEAR RESONANCE IN NON-PERFECT METALS: ALLOYS, LIQUID METALS, AND LIQUID ALLOYS

1. Introduction

IN this chapter we discuss several questions which at first sight seem unrelated. They all have one important feature in common: the translational symmetry of the lattice is to a certain extent broken. The problem of understanding nuclear resonance properties in an alloy is very important and provides interesting information about the metallic state. It is sometimes difficult in dilute alloys to discriminate against effects arising from the presence of other types of defects (point defects, dislocations, and so on). In both alloys and defects, large quadrupolar perturbations are observed; but for alloys there is also a magnetic effect coming from the perturbation induced in the conduction electron cloud by the foreign atom. The magnetic effects in alloys are discussed first.

2. Magnetic effects in alloys

2.1. *The Knight shift*

For a given nuclear moment the shift (neglecting for simplicity the non-contact hyperfine coupling) is proportional to the value of the electronic magnetization at its position. In a perfect metal all the average values of these local magnetizations are equal and, as shown in Chapter 4, the Knight shift is proportional to the space average electronic magnetization. The situation is different for an alloy. (We shall consider mainly the case of dilute alloys.) The nuclear moments in the vicinity of a foreign atom are submitted to a local field that differs from the field seen by nuclei far from the impurity and therefore they experience different values for the Knight shift.

The problem is to calculate the spatial variation of the electronic magnetization. Let us assume that in the alloy the electrons are described by a wave function of the form

$$\phi_{\mathbf{k}}(\mathbf{r}) = \tilde{U}_{\mathbf{k}}(\mathbf{r})e^{i\mathbf{k}.\mathbf{r}},$$

where the function $\tilde{U}_{\mathbf{k}}(\mathbf{r})$ is periodic and equal to the function $U_{\mathbf{k}}(\mathbf{r})$ far from the impurity; we also assume that a Fermi surface can still be defined. If the impurity is replaced by a potential a Fermi energy may still be defined. However as the lattice symmetry is broken, wave functions of different **k** values are admixed; we assume that the admixture far from the impurities remains small. The calculation of the Knight shift using these assumptions, at the point $\mathbf{R}_i$ is done exactly as in Chapter 4 leading to the result

$$K(\mathbf{R}_i) = \frac{8\pi}{3} V\chi'_{\mathrm{S}} |\phi_{k_{\mathrm{F}}}(\mathbf{R}_i)|^2. \qquad (7.1)$$

The problem is now to calculate the function $\phi_{\mathbf{k}}(\mathbf{r})$.

2.1.1. *Phase-shift methods.* We present first a calculation using a method due to Friedel.[83] Let us assume that only one foreign atom is present. We assume that the impurity produces a static potential which has to be added to the periodic potential. For simplicity this potential is assumed to have a spherical symmetry. The effect of this potential on the electronic wave functions is calculated using a phase-shift technique, a well-known method used in nuclear scattering theory.

In the absence of this potential V_{P} the wave functions are

$$\phi_{\mathbf{k}}^0(\mathbf{r}) = U_{\mathbf{k}}(\mathbf{r})\mathrm{e}^{\mathrm{i}\mathbf{k}\cdot\mathbf{r}}.$$

They are solutions of the Schrödinger equation in the presence of the pure periodic potential of the metal.

In the presence of V_{P} we look for a wave function of the form

$$\phi_{\mathbf{k}}(\mathbf{r}) = U_{\mathbf{k}}(\mathbf{r})F_{\mathbf{k}}(\mathbf{r}). \qquad (7.2)$$

Friedel has shown that for calculating $F_{\mathbf{k}}(\mathbf{r})$ it is sufficient to consider the effect of V_{P} over the plane-wave part $\mathrm{e}^{\mathrm{i}\mathbf{k}\cdot\mathbf{r}}$, neglecting the periodic potential (which is still taken into account by the presence of the function $U_{\mathbf{k}}(\mathbf{r})$ in eqn (7.2)).[82] The wave function $\mathrm{e}^{\mathrm{i}\mathbf{k}\cdot\mathbf{r}}$ is expanded in spherical partial waves around the impurity:

$$\mathrm{e}^{\mathrm{i}\mathbf{k}\cdot\mathbf{r}} = \sum_l \mathrm{i}^l(2l+1)j_l(kr)P_l(\cos\theta),$$

where j_l is the spherical Bessel function of order l, θ the angle between the vectors **k** and **r**, and P_l the Legendre polynomial of order l. If the effect of V_{P} is to produce a phase shift η_l for the lth order partial wave, the perturbed wave function that is a superposition of the incoming wave and of all the scattered partial waves is given by the equation

$$F_{\mathbf{k}}(\mathbf{r}) = \sum_l \mathrm{i}^l(2l+1)\,\mathrm{e}^{\mathrm{i}\eta_l}f_l(kr)P_l(\cos\theta), \qquad (7.3)$$

with

$$f_l(kr) = j_l(kr)\cos\eta_l - n_l(kr)\sin\eta_l.$$

n_l is the spherical Neumann function. Using this wave function the change in the electronic density for a given value of the wave vector $\mathbf{k}$, $\Delta P_{\mathbf{k}}(\mathbf{r})$ is obtained:

$$\Delta P_{\mathbf{k}}(\mathbf{r}) = |\phi_{\mathbf{k}}(\mathbf{r})|^2 - |\phi_{\mathbf{k}}^0(\mathbf{r})|^2$$
$$= |U_{\mathbf{k}}(\mathbf{r})|^2 \{|F_{\mathbf{k}}(\mathbf{r})|^2 - 1\};$$

using eqn (7.3) the relative change

$$\frac{\Delta P_{\mathbf{k}}(\mathbf{r})}{P_{\mathbf{k}}(\mathbf{r})} = \frac{\Delta P_{\mathbf{k}}(\mathbf{r})}{|U_{\mathbf{k}}(\mathbf{r})|^2}$$

is obtained.

If an average is taken over the direction of the wave vector $\mathbf{k}$ one finds the following result:

$$\frac{\Delta P_k(r)}{P_k(r)} = \sum_l (2l+1)[\sin^2\eta_l\{n_l^2(kr) - j_l^2(kr)\} - \sin 2\eta_l \, n_l(kr) j_l(kr)]. \quad (7.4)$$

At large distances from the impurity the asymptotic expansions of n_l and j_l may be used. We get

$$\frac{\Delta P_k(r)}{P_k(r)} = \sum_l (-1)^l (2l+1) \sin \eta_l \, \frac{\sin(2kr+\eta_l)}{(kr)^2}. \quad (7.5)$$

We notice the oscillating character of this function and the long range of the perturbation which decreases as $1/r^2$. The problem is solved if the set of all the phase shifts is known. We shall not discuss here the choice of the potential V_P from which we calculate the phase shifts.

It is possible to consider the phase shifts as parameters and verify if the various experimental data, such as the Knight shift, the resistivity, and the quadrupolar effects are accounted for by using a limited number of phase shifts.

There is an important relation between these phase shifts, usually called the Friedel sum rule, which is

$$Z = \frac{2}{\pi} \sum_l (2l+1)\eta_l(k_{\mathrm{F}}), \quad (7.6)$$

where Z is the difference in valency between the impurity and the host metal and $\eta_l(k_{\mathrm{F}})$ is the phase shift for electrons having the Fermi energy.[83] This relation strongly limits the possible values for the phase shifts.

This rule has the following physical origin. Due to the long-range Coulomb interaction, the electrons are attracted by the impurity if the nuclear charge is increased locally. The excess charge of the impurity is 'screened', that is to say the total change of charge integrated over a large volume around the impurity must be equal to zero. If this change is calculated using the wave function given by eqn (7.3) the Friedel sum rule is obtained. Finally, there is a correction to Z due to the fact that the impurity changes the volume of the cell around its site so that the change of charge per unit volume due to the impurity differs slightly from the difference between the valencies. Coming back to the Knight-shift evaluation the relative change is given by the relative change of the electronic density for electrons at the Fermi level and is given by eqn (7.4):

$$\frac{\Delta K(\mathbf{R}_i)}{K(\mathbf{R}_i)} = \frac{\Delta P_{k_F}(\mathbf{R}_i)}{P_{k_F}(\mathbf{R}_i)}.$$

If we consider a dilute alloy, for large dilutions there is no interference between the scattered waves coming from the different impurity centres, and the change of the Knight shift is obtained by adding the changes due to all the impurities.

In some cases the change of the Knight shift in the vicinity of the impurity is so large that a structure appears in the resonance line, but very often only a broadening of the line is observed (this broadening is proportional to the external field).

The centre of gravity of the line is also changed, the shift being given by the equation

$$\frac{\overline{\Delta K}}{K} = \frac{1}{N} \sum_i \frac{\Delta K(\mathbf{R}_i)}{K} = \frac{1}{N} \sum_{i,j} \frac{\Delta P_{k_F}(\mathbf{R}_{ij})}{P_{k_F}(\mathbf{R}_i)}, \tag{7.7}$$

where $\mathbf{R}_i$ is the position of the nucleus, $\mathbf{R}_j$ the position of an impurity, and N is the total number of nuclear spins. For dilute alloys, assuming a random distribution for the impurities, this equation becomes

$$\frac{\overline{\Delta K}}{K} = c \sum_i \frac{\Delta P_{k_F}(\mathbf{R}_i)}{P_{k_F}(\mathbf{R}_i)}, \tag{7.8}$$

where c is the concentration of impurities, the summation is taken over the nuclear sites, and $\Delta P_{k_F}(\mathbf{R}_i)$ is the variation of the density around an isolated impurity.

2.1.2. The perturbation theory approach. The phase-shift method is not the only possible approach to this problem. Another possibility is to consider the change induced in the wave function by the potential $V_P(r)$ using standard perturbation theory. As in the previous calculation only the influence of $V_P(r)$ on the plane-wave part of the wave function $\phi_k^0(\mathbf{r})$ will be considered. The perturbed function $F_k(\mathbf{r})$ (defined by eqn (7.2)) may be written[84]

$$F_k(\mathbf{r}) = \left\{ e^{i\mathbf{k}.\mathbf{r}} + \frac{2m}{\hbar^2} \sum_{|\mathbf{k}+\mathbf{q}|>k_F} \frac{\Phi(\mathbf{q})\, e^{i(\mathbf{k}+\mathbf{q}).\mathbf{r}}}{|\mathbf{k}|^2 - |\mathbf{k}+\mathbf{q}|^2} \right\}, \tag{7.9}$$

where $\Phi(\mathbf{q})$ is the matrix element of $V_P(r)$ between the plane wave functions $\mathbf{k}$ and $\mathbf{k}+\mathbf{q}$, or $\Phi(\mathbf{q})$ is the spatial Fourier transform of $V_P(r)$. This equation describes the lowest-order admixture of the wave functions.

However, we shall consider only the admixture with wave functions of empty electronic states. The potential $V_P(r)$ also produces an admixture with the wave functions of the occupied states, but such admixtures do not produce a change in the electronic densities, so these terms will be omitted,[84] using the wave function $F_\mathbf{k}(\mathbf{r})$ given by eqn (7.9). The relative change of the electronic density for electrons having the Fermi energy, keeping only the terms linear in $\Phi(\mathbf{q})$, is

$$\frac{\Delta P_{k_\mathrm{F}}(\mathbf{r})}{P_{k_\mathrm{F}}(\mathbf{r})} = \frac{2m}{\hbar^2} \sum_{|\mathbf{k}+\mathbf{q}|>k_F} \left\{ \Phi(\mathbf{q}) \frac{(e^{i\mathbf{q}\cdot\mathbf{r}}+e^{-i\mathbf{q}\cdot\mathbf{r}})}{|\mathbf{k}|^2-|\mathbf{k}+\mathbf{q}|^2} \right\}, \tag{7.10}$$

where an average over the directions of the vectors $\mathbf{k}$ assumed to be at the Fermi surface has to be taken. After performing this average one finds

$$\frac{\Delta P_{k_\mathrm{F}}(r)}{P_{k_\mathrm{F}}(r)} = \frac{2m}{\hbar^2} \sum_{\mathbf{q}} \Phi(q) \frac{q}{k_F} \ln\left|\frac{2k_\mathrm{F}+q}{2k_\mathrm{F}-q}\right| \{e^{i\mathbf{q}\cdot\mathbf{r}}+e^{-i\mathbf{q}\cdot\mathbf{r}}\}. \tag{7.11}$$

We notice that the quantity that has to be integrated presents a similarity with the function $\chi_S(q)$ defined in Chapter 3, with a singular derivative at $q = 2k_\mathrm{F}$. On integrating over the variable q this singularity produces the spatial oscillations of the density. Using eqn (7.11) the average Knight shift is easily estimated to be

$$\frac{\overline{\Delta K}}{K} = c \sum_{Q} \Phi(Q) \frac{Q}{2k_\mathrm{F}} \ln\left|\frac{2k_\mathrm{F}+Q}{2k_\mathrm{F}-Q}\right|, \tag{7.12}$$

the summation being taken only over the reciprocal lattice vectors $\mathbf{Q}$.

2.1.3. *Experimental results.* Many experimental studies of these effects have been made by choosing a solvent metal of nuclear spin $I = \frac{1}{2}$ to avoid quadrupolar effects. The first experiments were performed on silver alloys.[85] Generally speaking it is possible to explain the Knight-shift variation using a reasonable choice of phase shifts. The variation with concentration is found to be linear up to a concentration of about 10 per cent and sometimes the linear dependence is valid over an even larger range of concentration.[86]

In a rather qualitative fashion, the shifts increase if Z is increased in accordance with the prediction of the Friedel sum rule. Some results are presented in Fig. 7.1. In some cases it is possible to measure the Knight shift for the impurity nuclear moment when the concentration

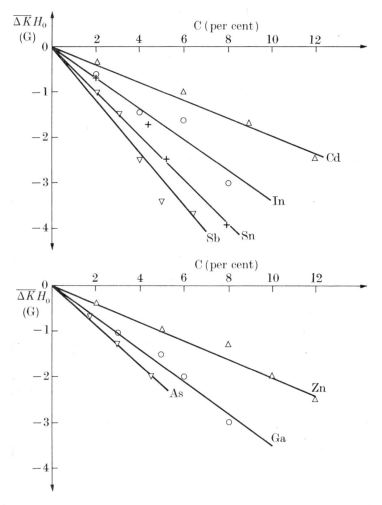

Fig. 7.1. Variation of the silver Knight shifts as a function of the impurity concentration.[86]

is sufficiently high. This shift has no reason to be related in a simple way to the shift of the pure solute metal. Even if the wave function keeps some similarity to the wave function of the free solute metal, the electronic susceptibility is related to the susceptibility of the host metal.

2.2. Broadening

As we know, the spatial variation of the Knight shift, the broadening of the line, may be calculated using a moment method. If we call ΔH_i the change of the local field at the point i due to the presence of impurities, the second moment of the line is given by the equation

$$M_2 = \frac{1}{N} \left\{ \sum_i (\Delta H_i^2) \right\} - \left(\frac{1}{N} \sum_i \Delta H_i \right)^2. \tag{7.13}$$

The second term on the right-hand side of the equation takes into account the change of position of the centre of the line. Let us compare the square root of this moment with the value of the Knight shift for the pure metal:

$$\frac{M_2^{\frac{1}{2}}}{KH_0} = \left[\frac{1}{N} \sum_{i,j} \left\{ \frac{\Delta P_{k_F}(\mathbf{R}_{ij})}{P_{k_F}(\mathbf{R}_i)} \right\}^2 - \left\{ \frac{1}{N} \sum_{i,j} \frac{\Delta P_{k_F}(\mathbf{R}_{ij})}{P_{k_F}(\mathbf{R}_i)} \right\}^2 \right]^{\frac{1}{2}}. \tag{7.14}$$

The first term is proportional to the concentration c, whereas the second varies as c^2 and is negligible for dilute alloys.

If this width is larger than the intrinsic width, a variation of the line width with the concentration varying as $c^{\frac{1}{2}}$ will be observed.

2.3. *The measurement of indirect interactions in alloys*

It is worth while to describe now an interesting experiment which takes advantage of the spatial variation of the Knight shift in an alloy.[87] We noticed in Chapter 4 that in a pure metal with only one type of nuclear spin the indirect scalar interaction does not contribute to the second moment and therefore its measurement is difficult. In a dilute alloy the nuclear spins at different positions experience different fields. Let us write the Hamiltonian for two nuclear spins:

$$\mathscr{H} = -\hbar\gamma_n H_0(I_{iz}+I_{jz}) - \hbar\gamma_n(\Delta H_i I_{iz} + \Delta H_j I_{jz}) + A_{ij}\,\mathbf{I}_i.\mathbf{I}_j \tag{7.15}$$

or, using the following equivalent form,

$$\mathscr{H} = \hbar\omega'_n(I_{iz}+I_{jz}) + \hbar\delta(I_{iz}-I_{jz}) + A_{ij}\,\mathbf{I}_i.\mathbf{I}_j,$$

where δ is proportional to $\Delta H_i - \Delta H_j$.

If $\delta = 0$ a single resonance line is observed and the scalar terms produce no effect, but when $\delta \gg A_{ij}$ the spins are resonating at different frequencies and the exchange term produces a splitting of the resonance lines. In the lowest order of perturbation theory the following Hamiltonian has to be diagonalized:

$$\mathscr{H} = \hbar\omega'_n(I_{iz}+I_{jz}) + \hbar\delta(I_{iz}-I_{jz}) + A_{ij}\,I_{iz}\,I_{jz}.$$

The quantity A_{ij} may be measured (the absorption spectrum is shown on Fig. 7.2).

In alloys a large number of nuclear spins are present, the resonance lines are not resolved, and only a broadening is observed. However, it will be recalled from Chapter 1 that such an inhomogeneous broadening

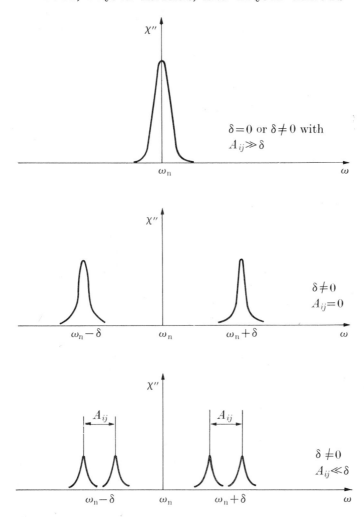

FIG. 7.2. The resonance spectrum of the Hamiltonian given by eqn (7.15).

may be suppressed using a spin-echo method. It is thus possible to measure the damping due to the $A_{ij} I_{iz} I_{jz}$ terms.[72,87]

This method has been used for measuring the indirect interactions in several metals, particularly in platinum, lead,[88] cadmium, and tin.[72]

2.4. *The effect of alloying on indirect interactions and relaxation rates*

A possible objection to the experimental technique just described for measuring the indirect interactions is that we may suspect that the alloying changes the value of the indirect coupling from the value in the pure metal. Using modified wave functions, let us calculate in the alloy the indirect interaction between the spins i and j. We find the change in A_{ij} to be given by the equation

$$\frac{\Delta A_{ij}}{A_{ij}} = \frac{\Delta P_{k_{\mathrm{F}}}(\mathbf{R}_i)}{P_{k_{\mathrm{F}}}(\mathbf{R}_i)} + \frac{\Delta P_{k_{\mathrm{F}}}(\mathbf{R}_j)}{P_{k_{\mathrm{F}}}(\mathbf{R}_j)}.$$

This change is simply the sum of the relative changes of the Knight shift for the two spins

$$\frac{\Delta A_{ij}}{A_{ij}} = \frac{\Delta K(\mathbf{R}_i)}{K} + \frac{\Delta K(\mathbf{R}_j)}{K}. \tag{7.16}$$

From the measurement of the broadening of the line an estimate of the change of the Knight shift is obtained and therefore an estimate of the change in the indirect interactions. The method for measuring A_{ij} previously described is valid if

$$A_{ij} \ll \Delta H_i - \Delta H_j \quad \text{or} \quad A_{ij} \ll \{\Delta K(\mathbf{R}_i) - \Delta K(\mathbf{R}_j)\}H_0.$$

On the other hand, to have a measurement of A_{ij} not perturbed by the alloying we must satisfy the condition $\Delta A_{ij} \ll A_{ij}$. Using (7.16), these two conditions imply

$$\Delta A_{ij} \ll H_0\{\Delta K(\mathbf{R}_i) - \Delta K(\mathbf{R}_j)\},$$

or $$A_{ij}\{\Delta K(\mathbf{R}_i) + \Delta K(\mathbf{R}_j)\} \ll KH_0\{\Delta K(\mathbf{R}_i) - \Delta K(\mathbf{R}_j)\}.$$

If we assume that only one of the Knight shifts is changed, this condition means simply that the indirect interaction must be smaller than the Knight shift (this condition is always fulfilled).

There is also a change in the relaxation rate that becomes a function of the position of the nuclear spin. The correction is given by the equation

$$\Delta\left\{\frac{1}{T_1}(R_i)\right\} \bigg/ \frac{1}{T_1} = 2\frac{\Delta K(R_i)}{K}, \tag{7.17}$$

provided the relative change in the Knight shift is small.

3. Quadrupolar effects

We shall assume that the symmetry around a nuclear spin in the perfect metal is cubic. In the vicinity of a defect an electric field gradient is produced. The experimental results are analysed, taking into account the relative magnitude of the quadrupolar splitting and of the nuclear line width.

3.1. *The first-order effect*

When for almost all the nuclear spins the quadrupolar splitting is such that the satellite lines are displaced outside the unperturbed line shape, it is said that we have a first-order quadrupolar effect and only the central component is observed.

In this case one observes a change in the intensity of the nuclear resonance signal. The reduction in intensity is a function of the value of the spin I; for a spin $I = \frac{3}{2}$ the intensity of the central component is 40 per cent of the total intensity. More generally, the reduction factor ρ is given by the relation

$$\rho = |\langle \tfrac{1}{2}|I_+|-\tfrac{1}{2}\rangle|^2 \Big/ \sum_{m=I}^{m=1-I} |\langle m|I_+|m-1\rangle|^2, \qquad (7.18)$$

which gives the following result:

$$\rho = \frac{3}{2}\frac{(I+\tfrac{1}{2})^2}{I(I+1)(2I+1)} = \frac{3}{8}\frac{(2I+1)}{I(I+1)}. \qquad (7.19)$$

It is noticed that ρ decreases when the value of the spin I increases. If some spins are submitted to a splitting smaller than the line width, the value of the observed ρ will fall between 1 and the value given by eqn (7.19).

3.2. *The second-order effect*

If to well-annealed copper one adds another metal the satellite lines disappear for concentrations of less than 1 per cent; and if the concentration is further increased the intensity of the central line itself decreases rapidly as a function of the concentration and no broadening of the line appears. It is said that a second-order quadrupole effect is observed.

This experimental behaviour is explained by a simple model.[29] Let us assume that around each impurity a radius r_c can be defined such that inside the sphere of radius r_c centred around the impurity the nuclear moments are submitted to a gradient strong enough to render the line (the central component) unobservable; outside the sphere the spins are not affected. With this model, knowing the variation of the intensity as a function of the concentration, r_c may be estimated. Let us assume that n nuclear spins are present inside the sphere of radius r_c. The intensity will be proportional to the probability for a given spin to find around it n sites without an impurity: this probability is $(1-c)^n$ if the impurities are distributed randomly. If the logarithm of the intensity is plotted versus c a straight line is obtained the slope of which measures n (Fig. 7.3).

This model is justified by the fact that the field gradient around the impurity varies like $1/r^3$ (this result will be proved later) and the broadening of the central component therefore varies as $1/r^6$. This fast variation

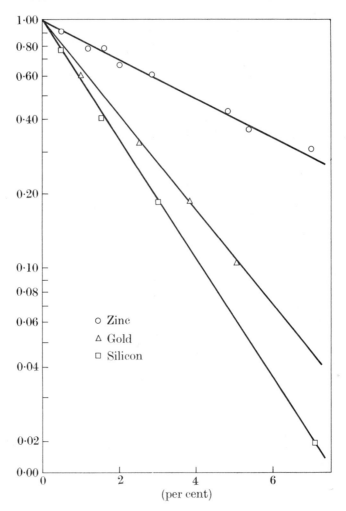

FIG. 7.3. Intensity of the copper resonance signal as a function of the impurity concentration.[90]

explains why a nuclear spin experiences a large quadrupolar effect or no effect.

The same analysis with an 'all or nothing' effect is sometimes also applied for the first-order effect. Here also $\log I$ is plotted versus $1-c$. Of course, because the first-order effect varies only as $1/r^3$, the validity of this model is more questionable. This situation is found in aluminium alloys. In Table 7.1 we give some values of n, called the 'wipe-out number'.

However, the result may be correctly interpreted only if the impurities are distributed in a random way. It has been proved that in some samples the annealing process may change the spatial repartition of the impurities. In such a case the variation of I with the concentration does not give the correct value for n (see, for instance, Ref. 91).

TABLE 7.1

Some values of n in copper and aluminium alloys. The value n_1 for aluminium corresponds to a first-order effect, whereas n_2 for copper is for a second-order effect

Solute	Al	Ref.	Cu	Ref.
Zn	$n_1 = 98$	90	$n_2 = 18$	29
Mg	$n_1 = 130$	90		
In			$n_2 = 46$	89
Si			$n_2 = 58$	89
Au			$n_2 = 46$	89

3.3. *Direct observation of quadrupolar splitting in alloys*

Measurement of the variation of the intensity of the line does not allow a direct determination of the quadrupolar splitting near the impurity but gives only the number of spins submitted to a gradient of field larger than a certain value. It is quite difficult to measure the absorption signal for these spins directly because of the lack of sensitivity (although a direct absorption signal was detected in an Al–Mg alloy using an averaging method (Ref. 92)). Redfield has proposed a technique[93] for measuring these splittings. Let us briefly discuss the method and the results so obtained. The sample (these experiments have been performed on copper or aluminium alloys) is in a high external magnetic field. The field is suppressed in a time short compared to the relaxation time. The experiments are done at low temperatures in order to have a sufficiently long relaxation time. If the system is described by a spin temperature, this demagnetizing process reduces that temperature. If the field is raised again towards its initial value a resonance signal can be observed whose magnitude is proportional to the nuclear equilibrium magnetization in the initial static field. Now if during the period of time when no applied field is present a radiofrequency whose frequency is equal to a quadrupolar splitting for some spins is applied, these spins will absorb energy, their temperature will increase, and this heating is partially transmitted to the whole nuclear spin system. If the field is raised again the observed signal is weaker because the spin temperature is higher.

By varying the value of the frequency several resonances are observed corresponding to nuclear spin nearest neighbours, next nearest neighbours, and so on. The results are given in Table 7.2 and will be discussed after the theoretical calculation. The variation of an aluminium signal as a function of the frequency of the heating field is presented in Fig. 7.4.

TABLE 7.2

Quadrupolar frequencies in copper alloys

Shell	Ref.	1	2	3	4
Number of atoms in the shell		12	6	24	12
Ag ν (kHz)	93	620	475	200	75
theory	95	380	155	82	13
Zn ν (kHz)	93	> 5000	1980	90	
theory	95	650	290	180	8
	96	610	420	70	

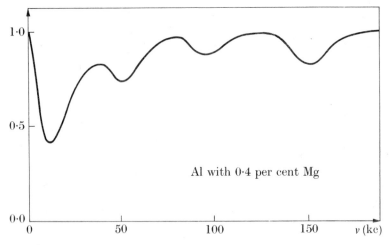

Al with 0·4 per cent Mg

Fig. 7.4. Relative variation of the aluminium signal in high field as a function of the frequency of the heating field. The peak at very low frequency is not due to a quadrupolar splitting but to the heating of the spin-spin energy of the aluminium nuclei.[91]

3.4. *Calculation of quadrupolar effects*

The first suggestion for explaining the magnitude of the quadrupolar effects in a dilute alloy was to consider the influence of the distortion of the lattice due to the presence of the impurity. This point of view was justified by the well-known fact that the main part of the quadrupole interactions in a metal has an ionic origin. That this explanation is not correct was shown experimentally by Averbuch *et al.*[94] Using X-rays, they measured the amount of deformation in a crystal of cold-worked copper. The electric field gradient coming from these measured deformations was estimated and found to be much too small to explain the observed change in the intensity of the line. The quadrupolar effects arise from the deformation of the electronic cloud. Several calculations have been made for dilute alloys.[95,96] Friedel's calculation, which is very similar to the calculation of the change in the Knight shift, will be given. Friedel uses the wave functions defined by eqns (7.2) and (7.4). If the distance to the impurity is large enough, an asymptotic expansion of $F_{\mathbf{k}}(\mathbf{r})$ is used:

$$F_{\mathbf{k}}(\mathbf{r}) = \sum_l P_l(\cos\theta) \, i^l (2l+1) \frac{1}{kr} \sin\left(kr + \eta_l - \frac{l\pi}{2}\right).$$

From the knowledge of $F_{\mathbf{k}}(\mathbf{r})$ the change $\Delta P_{\mathbf{k}}(\mathbf{r})$ in the electronic density at the point $\mathbf{r}$ for electrons with wave vector $\mathbf{k}$ is calculated. The total change is obtained by integrating $\Delta P_{\mathbf{k}}(\mathbf{r})$ over all the wave vectors inside the Fermi sphere:

$$\Delta P(\mathbf{r}) = \int_0^{k_F} d^3k \, \Delta P_{\mathbf{k}}(\mathbf{r}).$$

The result is

$$\Delta P(\mathbf{r}) = -\sum_l (-1)^l \frac{(2l+1)}{4\pi^2 r^3} \sin \eta_l^F \{u_{k_F}^2(r) \, e^{i(2k_F r + \eta_l)} + \text{complex conjugate}\}. \tag{7.20}$$

For free electrons this becomes

$$\Delta P(\mathbf{r}) = -\sum_l (-1)^l \frac{(2l+1)}{2\pi^2 r^3} \sin \eta_l^F \cos(2k_F r + \eta_l^F), \tag{7.21}$$

which is written by Friedel as

$$\Delta P(\mathbf{r}) = -\frac{\alpha}{2\pi^2 r^3} \cos(2k_F r + \phi). \tag{7.22}$$

α and ϕ are calculated as functions of the phase shifts:

$$\left.\begin{aligned} \alpha \cos\phi &= \tfrac{1}{2} \sum_l (-1)^l (2l+1) \sin 2\eta_l^F \\ \alpha \sin\phi &= \sum_l (-1)^l (2l+1) \sin^2 \eta_l^F \end{aligned}\right\}. \tag{7.23}$$

To calculate the field gradient at the point R_i (the impurity being at the origin) we have to consider two contributions, one from the excess ionic charge located at the origin, the other from the electrons whose distribution is described by eqn (7.20). If we assume this gradient to have

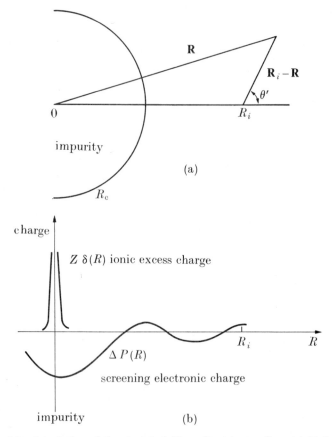

FIG. 7.5. Calculation of the electric field gradient in an alloy. (a) Definition of the vector $\mathbf{R}$ and angle θ'. (b) Variation of the electric charges on the axis joining the impurity to the nucleus i.

only a component along the vector $\mathbf{R}_i$, it has the value

$$G(\mathbf{R}_i) = -2 \int [\Delta P(\mathbf{r}) - Z\delta(\mathbf{r})] \frac{P_2(\cos \theta')}{|\mathbf{r}-\mathbf{R}_i|^3} \, d^3r, \qquad (7.24)$$

where θ' is the angle between the vectors $\mathbf{R}_i$ and $\mathbf{r}-\mathbf{R}_i$ (see Fig. 7.5).

As we remarked earlier, the electronic charges $\Delta P(\mathbf{r})$ screen the excess ionic charge Z, and if one considers a sphere around the impurity the radius R_c of which is of the order of the unit cell radius the following relation is obtained:

$$Z = \int\limits_{R < R_c} \Delta P(\mathbf{r}) \, d^3r;$$

and in first approximation the charges inside the sphere do not contribute to the gradient given by eqn (7.24). The largest contribution is obtained when r is in the vicinity of $\mathbf{R}_i$ and in this region $\Delta P(\mathbf{r})$ may be replaced by its asymptotic expansion (7.20); if the integration is rather arbitrarily limited to the Wigner–Seitz sphere around $\mathbf{R}_i$ we have

$$G(R_i) = \frac{8\pi}{3} \frac{\alpha}{2\pi^2} \mu \frac{\cos(2k_{\mathrm{F}} R_i + \phi)}{R_i^3},$$

where the quantities α and ϕ have already been defined by eqn (7.23) and

$$\mu = \frac{3}{8\pi} \int\limits_{\text{Wigner–Seitz}} e^{2i\mathbf{k}_{\mathrm{F}}^i \cdot \mathbf{r}} \{U_{\mathbf{k}_{\mathrm{F}}^i}(\mathbf{r})\}^2 \frac{P_2(\cos\theta)}{r^3} \, \mathrm{d}^3 r. \qquad (7.25)$$

$\mathbf{k}_{\mathrm{F}}^i$ is the Fermi wave vector along the direction $\mathbf{R}_i$. The advantage of this notation is that it can be shown that the quantity μ is equal to one for free electrons and so this factor gives the influence of the band structure on the field gradient.[95]

TABLE 7.3

Quadrupolar effects in aluminium alloys

Shell		Ref.	1	2	3
Mg	theory	96, 98	48	27	12
($\Delta Z = -1$)	experiment	97, 98, 91	185	63	< 15
Zn	theory	96, 98	79	6	6
($\Delta Z = -1$)	experiment	97, 98, 91	180	35	< 15
Ag	theory (a)	96, 98	185	12	0
($\Delta Z = -2$)	(b)		160	36	12
	experiment	98, 91	215	52	

The figures are the field gradient in units of 10^{-21} cm³. In the calculations μ has the value $\mu = 12$. The two theoretical values for silver are obtained using two possible sets of phase shifts.

We notice several differences with the Knight-shift calculation. First, as we are dealing with a charge effect all the values of $\mathbf{k}$ contribute. Then in the final result we find two types of parameters, α, ϕ, which are functions of the phase shifts, and μ, which is a function of the band structure. This last coefficient is not known with great accuracy and did not appear in the Knight-shift calculation.

Table 7.2 shows the results obtained by Redfield in copper alloys together with two theoretical values. In Table 7.3 the results in aluminium alloys are given. For this metal there is an additional complication because the spin $I = \frac{5}{2}$ and there are two resonant frequencies for a given spin.

In aluminium the theoretical values are obtained by assuming that only two phase shifts are important, η_0 $(l = 0)$ and η_1 $(l = 1)$. These phase shifts are deduced from an equation giving the part of the resistivity due to the impurity as a function of the phase shifts

$$\Delta\rho = \frac{4\pi\hbar c}{Z_A k_F e^2} \sum_l l \sin^2(\eta^F_{l-1} - \eta^F_l)$$

(where Z_A is the valency of the pure metal and $\Delta\rho$ the increase in resistivity due to the alloying) and from the Friedel sum rule.

The observed gradients are usually larger than expected. This discrepancy may be a consequence of the constant use in the theory of asymptotic expansions for the wave functions, whereas the experiments measure the gradient in the vicinity of the impurity where such an expansion may be not valid.

4. Resonance in liquid metals and liquid alloys

A large number of nuclear magnetic resonance experiments have been performed in liquid metals and liquid alloys.

Two motivations are found for this research. One, of purely experimental origin, is that in a liquid the motion of the atoms produces a narrowing of the resonance line and, instead of measuring a broadening due to the variation of the Knight shifts, we measure directly the average value of this quantity with very great accuracy. The other theoretical reason is that it is interesting to investigate the theoretical models of liquid metals. A liquid metal maintains its metallic character although there is no longer long-range order. The electrons move in a disordered lattice; crudely speaking, we may expect that the relation between the energy and the wave vector **k** will be more comparable to the relation in a free-electron model, because in a solid metal this relation is strongly affected by the existence of Brillouin zones. First, we shall discuss the changes in nuclear resonance properties produced by melting.

4.1. *Variation of the Knight shift at the melting-point*[99]

It is found that for many metals such as Li, Na, Rb, Cs, Hg, Al, and Sn the Knight shift changes at the melting-point by only a small amount. As an example, for Na there is a 2 per cent increase of the Knight shift (in other cases a decrease is observed). This is because in all these metals the Fermi wave vectors are rather far from the wave vector at the Brillouin zone boundary, and the electronic density and susceptibility are not changed very much by the melting. In the case of sodium the variation has been calculated; the wave functions of the solid phase are assumed to be plane waves perturbed by the periodic potential, and in the liquid phase[100] plane waves perturbed by a random potential. Good agreement is obtained.

For some metals large variations are observed: in the case of bismuth the very small isotropic shift in the solid becomes, in the liquid, comparable to the value found for lead (which is of 1·47 per cent). We have already noted (see Chapter 6, Fig. 6.3) the large change observed in cadmium. In these metals the Fermi surfaces in the solid phase are complicated and the density of states rather small; in the liquid phase the electrons behave like free electrons and a large increase in the density of states is expected, so explaining in a very qualitative fashion the large changes in these Knight shifts.

4.2. *Relaxation rates in liquid metals*[101,102]

In a liquid metal the relaxation rate has two origins. One, as in the solid phase, is a modulation of the hyperfine coupling by the electronic motion. This relaxation rate is proportional to $k_B T$ and is usually comparable to the value extrapolated from the measurement at low temperatures in the solid.

To these rates is added another contribution due to the ionic motion in the liquid. This motion can modulate the spin–spin coupling as in the diffusion region in the solid, but this rate is found to be usually negligible. The ionic motion can also modulate the quadrupolar interaction, which is more important because the quadrupolar interaction is much larger than the spin–spin interaction.

In a liquid metal the electronic wave functions are not known and it is not very convenient for calculating the electric field gradient to separate the ionic and the electronic parts as described in Chapter 2.

The simplest model is to consider the field due to the ion screened by the electronic cloud. At a given instant the electric gradient is the sum of the appropriate derivatives of the screened potential of the ions. We assume that the ionic charges obey in the liquid a diffusion model and the relaxation rate is estimated as in an ordinary liquid (see Ref. 1, chap. 7; Refs. 102, 103). The main uncertainty in this calculation is the value of the antishielding factor γ. If the symmetry in the solid phase is not cubic the static quadrupolar splitting is known and, using the calculated value of the gradient in the solid phase, the antishielding factor is deduced for use in calculating the relaxation rate.

As mentioned in Chapter 5, if there are two isotopes it is possible to separate the magnetic and quadrupolar rates. This situation is found in gallium (^{69}Ga and ^{71}Ga), rubidium, and antimony. In the case of gallium the Scholl theory,[102] using the screened potential, is in quite good agreement with experiment. If there is only one isotope (indium) it is possible to study the temperature dependence of the relaxation rate. The temperature dependence of the rate due to the electronic motion is well known so that the variation of the quadrupolar part may

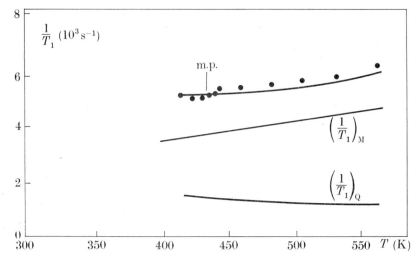

Fig. 7.6. Relaxation rate for In versus T. The curves $(1/T_1)_m$ and $(1/T_1)_Q$ are respectively estimates of the magnetic and quadrupolar contribution to the relaxation rate.[104] The required curve is obtained from the sum of these two contributions.

be estimated. It is found that this rate decreases slowly when the temperature increases. This effect is due to the increase in the diffusion rate. For indium the theory is in fair agreement with the experiments from 450 to 550 K (see Fig. 7.6).

4.3. *The Knight shift in liquid alloys*

There is a large number of systematic measurements of the variation of the Knight shift as a function of the composition in liquid alloys. The shift varies linearly with the concentration, c, over a very large range (at least for $c \leqslant 10$ per cent) (Fig. 7.7).

Let us call $\Delta K(\mathbf{R})$ the relative variation of the shift at the point R (the impurity is at the origin). Equation (7.7) valid in a solid must be written here as

$$\overline{\frac{\Delta K}{K}} = c \int \frac{\Delta K(\mathbf{R})}{K} \, P(\mathbf{R}) \, \mathrm{d}^3 R. \qquad (7.26)$$

$P(\mathbf{R})$ is the pair correlation function of the liquid, which gives the probability of finding a given atom at the point $\mathbf{R}$, assuming the presence of another given atom at the point $R = 0$. The function $P(\mathbf{R})$ may be deduced from studies by neutron or X-ray diffraction. For calculating

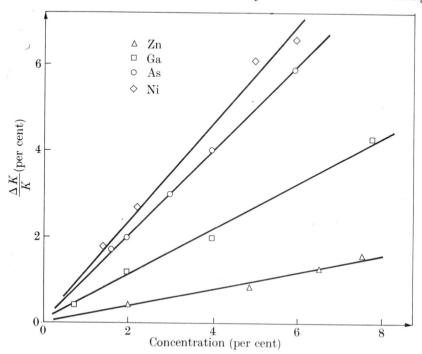

FIG. 7.7. Knight shifts in liquid copper alloy as a function of impurity concentration.[105]

$\Delta K(\mathbf{R})$, it is not a good approximation to use a phase-shift technique for the following reason: in a solid $P(\mathbf{R})$ is simply a sum of delta functions

$$P(\mathbf{R}) \sim \sum_{i \neq 0} \delta(\mathbf{R} - \mathbf{R}_i),$$

where $\mathbf{R}_i$ is equal to a lattice translation; in a liquid this function has the shape shown in Fig. 7.8. We notice that when R goes towards zero, P is small but finite and in this region $\Delta K(\mathbf{R})$ is large. The largest contribution to the integral (7.26) is due to small values of R, where, unfortunately, the phase-shift technique cannot be applied. It may be more convenient to use the perturbation approach. The generalization of eqn (7.12) for a liquid is

$$\overline{\frac{\Delta K}{K}} \sim c \int \Phi(\mathbf{q})\{1 - a(\mathbf{q})\}\frac{q}{k_\mathrm{F}} \ln\left|\frac{2k_\mathrm{F} + q}{2k_\mathrm{F} - q}\right| \mathrm{d}^3 q. \qquad (7.27)$$

$1-a(\mathbf{q})$ is the Fourier transform of the pair correlation function $P(\mathbf{r})$. In principle, this equation holds no more physical information than is obtained by the phase-shift analysis. The advantage is that instead of

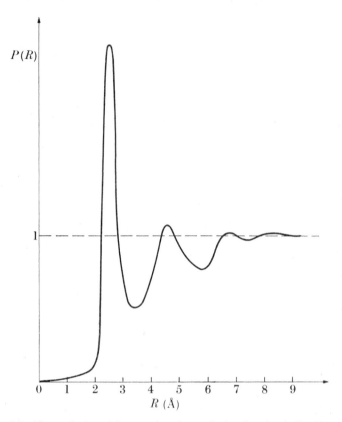

FIG. 7.8. The variation of the atomic pair correlation function in liquid copper.

considering $\Phi(\mathbf{q})$ as being the Fourier transform of $V_{\mathrm{P}}(r)$, this function may be considered as an effective 'pseudo-potential' that can be measured using the experimental data. However, as the function $1-a(\mathbf{q})$ is oscillating, we have to integrate very accurately for large values of q in eqn (7.27) and usually this technique does not produce significantly better results than the phase-shift treatment.[105]

8

NUCLEAR RESONANCE IN METALS AND ALLOYS WITH TRANSITION ELEMENTS

1. Introduction[106,107]

THE transition elements are those with unfilled internal shells. There are three groups having incomplete d-shells, viz. the iron group ($3d$), the palladium group ($4d$), and the platinum group ($5d$). There are also the rare-earth group with an unfilled $4f$-shell and, finally, the actinide group ($5f$- or $6d$-shells unfilled). A very large number of nuclear resonance experiments have been performed recently on transition metals or alloys; we may even say that during the last few years the largest part of the nuclear resonance studies in metals has been done in transition metals or alloys.

The magnetic behaviour in these metals differs from that described for 'simple metals' and is more complex. There is a variety of different behaviours, a fact that strongly suggests that no unique model will be able to explain all the experimental features. We shall now present two extreme models.

For certain non-magnetic metals of the iron group or of the $4d$- and $5d$-groups, such as vanadium, molybdenum, and tungsten, the metal can be described as having two overlapping energy bands; the Fermi level crosses the two bands. One of these bands is very broad and is similar to the bands found in simple metals (the wave function has a large s-character), the other is narrower and the density of states at the Fermi level is consequently larger; its wave functions present a large d-character.

A completely different situation is found for rare-earth metals (in their paramagnetic state at high temperature). In these metals there is still a broad s-band, but the f-electrons are 'localized'. By 'localized' we describe the situation for electrons that behave as f- (or d-) electrons in an ion located in an ionic crystal. Of course this definition is rather extreme; as we shall discuss later, the 'localized' f-electrons interact with the non-localized conduction electrons and consequently present some properties of non-localized electrons.

Between these two extreme models there are many possible inter-mediate situations.

The case of alloys is also difficult. There the basic problem of under-standing the behaviour of one magnetic impurity interacting with the conduction electron cloud has received a considerable number of theoretical treatments, leading sometimes to different conclusions. Even the experimental situation in alloys is far from being completely clear and only results believed to be well understood (at least by the author) will be discussed. In some cases the results (of nuclear resonance or other experiments) may be interpreted using rather different models.

2. Metals and alloys in a d-band model

2.1. *Susceptibility and Knight shift*

In transition metals, even more than for simple metals, the nuclear resonance cannot provide sufficient information by itself, and it is essential to know about other experimental properties such as the specific heat, susceptibility, and so on.

The results are analysed by assuming that the electrons are not interacting and occupy the two bands having, respectively, the density of states $g_s(E_F)$ and $g_d(E_F)$.

Let us discuss the various contributions to the static susceptibility and to the Knight shift. The susceptibility χ as measured by a static experiment is usually written as

$$\chi = \chi_s + \chi_d + \chi_{VV} + \chi_L + \chi_{dia}, \tag{8.1}$$

where χ_s and χ_d are, respectively, the spin susceptibilities for the electrons in the s- and d-bands. Using eqn (3.19),

$$\chi_s = \tfrac{1}{4}(g\beta)^2 g_s(E_F) \quad \text{and} \quad \chi_d = \tfrac{1}{4}(g\beta)^2 g_d(E_F) \tag{8.2}$$

(the values are given for a unit volume). Because $g_d(E_F)$ has a large value, χ_d is the largest contribution. χ_{VV} (where the symbol VV means Van Vleck) is an orbital contribution to the susceptibility. In Chapter 4 in the calculation of the orbital contribution to the Knight shift, we remarked that the coupling between the external field and the total orbital moment admixes the wave functions having the same value of k. In transition metals the structure of the d-band is rather complex and one finds several d-wave-functions for a given k value. As the band is narrow the energies of the d-levels do not differ very much and the admixture is rather large. There is also an admixture between the s- and d-wave-functions. These admixtures have the effect not only of producing an average orbital hyperfine field, but also of giving an average total orbital moment. If the susceptibility χ_{VV} is defined by

$$\chi_{VV} = \frac{\beta}{H_0} \langle \mathbf{L} \rangle,$$

where $\mathbf{L}$ is defined by eqn (4.18), using the periodic gauge (4.21) the calculation becomes very similar to the Knight-shift calculation, and the result is

$$\chi_{VV} = \beta^2 \sum_{n,n',\mathbf{k}} \frac{{}^2L_{\mathbf{k}n,\mathbf{k}n',z}\,{}^2L_{\mathbf{k}n',\mathbf{k}n,z}\,n_{\mathbf{k}n}(1-n_{\mathbf{k}n'})}{E_{\mathbf{k}n'} - E_{\mathbf{k}n}} + \text{complex conjugate.}$$

(8.3)

${}^2L_{\mathbf{k}'n',\mathbf{k}n,z}$ is defined by eqn (4.18′). In eqn (8.3) the largest contribution is given when both the energies E_{kn} and $E_{kn'}$ are energies in the d-band. The last two contributions are written for the sake of completeness but they are usually very small. χ_L is the Landau diamagnetic susceptibility and χ_{dia} the ionic susceptibility. As a matter of fact the separation of the orbital part into two terms χ_L and χ_{VV} is rather arbitrary and again a change in gauge changes each term, leaving their sum constant. With our choice, χ_{VV} given by eqn (8.3) is large and χ_L may be omitted. We may break up the Knight shift into its three largest contributions as follows:

$$K = K_s + K_0 + K_d.$$

(8.4)

The first term K_s comes from the contact interaction with the electrons in the s-band and is proportional to χ_s. K_0 is the orbital contribution to the Knight shift, which was calculated in Chapter 4, eqns (4.17) and (4.20). Here again, with our choice of gauge, the contribution (4.20) is the largest and the other will be omitted. We have not included the dipolar contribution due to the electrons in the s-band (the s-band has always a noticeable p-character) as it is usually smaller than K_s. Finally, let us consider the local field produced by the d-electrons. A wave function having purely d-character does not produce a contact hyperfine field, but as discussed in Chapter 2, the d-electrons are able to polarize the inner s-shells (and also the conduction s-electrons), thereby giving a contribution to the Knight shift K_d. This contribution is proportional to the number of unpaired d-electrons and therefore to χ_d. This effect is discussed further in Appendix 2. For sites with non-cubic symmetry there is also the possibility of a dipolar field due to the d-electrons.

There is an important consequence of the narrowness of the d-band, or more precisely of the fact that the function $g_d(E)$ varies appreciably over narrow energy ranges; the susceptibility χ_d varies with the temperature. This quantity is given by the relation (3.18), which may be written as

$$\chi_d \simeq \int_0^\infty \frac{\mathrm{d}n(E)}{\mathrm{d}E} \, g(E) \, \mathrm{d}E.$$

There are two reasons for the temperature dependence.[108] The first can be seen from the fact that $\mathrm{d}n(E)/\mathrm{d}E$ is a function of the energy having a width of the order of $k_B T$ but keeping a constant area. If $g(E)$ is written as

$$g(E) = g(E_F) + (E - E_F) g'(E_F) + \tfrac{1}{2}(E - E_F)^2 g''(E_F) + \cdots$$

the third term of this expansion gives a contribution to the susceptibility varying as $g''(E_F)(k_B T)^2$. The second term does not contribute because $\mathrm{d}n/\mathrm{d}E$ is an even function of $(E - E_F)$.

Another effect arises because E_F itself is a function of the temperature. This energy is calculated using the implicit equation

$$N = \int_0^\infty n(E) g(E) \, \mathrm{d}E;$$

integrating by parts we get

$$N = -\int_0^\infty \frac{\mathrm{d}n}{\mathrm{d}E} H(E) \, \mathrm{d}E, \quad \text{with} \quad H(E) = \int_0^E g(E') \, \mathrm{d}E'.$$

Using the same method $H(E)$ is expanded as a function of $E - E_F$:

$$H(E) = \int_0^{E_F} g(E) \, \mathrm{d}E + (E - E_F) g(E_F) + \tfrac{1}{2}(E - E_F)^2 g'(E_F) + \cdots$$

and

$$N = \int_0^{E_F} g(E) \, \mathrm{d}E + \frac{\pi^2}{6} (k_B T)^2 g'(E_F).$$

If E_F^0 is the Fermi energy for $T = 0$,

$$N = \int_0^{E_F^0} g(E) \, \mathrm{d}E.$$

Subtracting the two equations the change in the Fermi energy is obtained:

$$E_F = E_F^0 - \frac{\pi^2}{6} (k_B T)^2 \frac{g'(E_F^0)}{g(E_F^0)}.$$

The combined variation of χ at low temperature is therefore

$$\chi_d(T) = \chi_d(0) \left[1 + \frac{\pi^2}{6} (k_B T)^2 \left\{ \frac{g''(E_F^0)}{g(E_F^0)} - \left(\frac{g'(E_F^0)}{g(E_F^0)} \right)^2 \right\} \right]. \tag{8.5}$$

This equation is valid provided $k_B T$ is not large compared to the range of energy giving rise to changes in $g(E)$. If the susceptibility varies with the temperature, the part K_d of the Knight shift also varies. By comparing the change with temperature of K and χ it is in some cases possible to separate out the contribution K_d.

The term K_d has no reason to have the same sign as the Knight shift in a simple metal. Experimentally it is found that for some transition metals the total Knight shift K is negative. This result cannot be explained if the core polarization contribution K_d is omitted.

In all these considerations the effect of the interaction between the electrons is taken into account only partially (indeed the core-polarization effect is due to the exchange interaction between electrons) and this simplified treatment is seriously questionable for some metals. In the part devoted to the discussion of experimental results it will be shown that for some metals the susceptibilities are greatly enhanced by the electron–electron interactions.

2.2. *The relaxation rate*

The calculation is done as in Chapter 4. Several contributions of comparable magnitude are found. The s-band electrons give a contribution due to the contact interaction. The d-band electrons give contributions due to the orbital and dipolar interactions. There is also a possible contribution due to the matrix elements of the dipolar and orbital terms between the wave functions of the two bands. Finally, there is a relaxation rate due to the core-polarization effect.

Among all the processes the three rates that are usually predominant are the contact rate for the s-electrons and the orbital and core polarization rates for the d-electrons. The contact rate is large because the electronic density for s-electrons at the nuclear position is large. The two other processes are large because the density of states for d-electrons is large. The calculation of the contact and orbital rates is given in Chapter 4. The calculation of the last contribution requires some comment. It is, of course, not possible to calculate this rate in a model neglecting the interactions between electrons; this term exists only because the d-electrons interact with the inner (or outer) s-electrons. A detailed calculation is given in Reference 109 and will be included in Appendix 2. The principle of the calculation is as follows:

First an N-electron wave function ψ_0 is built that includes the s-electron in the internal shell of the atoms. Then one considers the effect of the Coulomb interaction that admixes different configurations.

Yafet and Jaccarino[109] use tight binding wave functions for the d-electrons. With these perturbed N-electron wave functions they calculate the matrix elements of the hyperfine interaction and obtain the Knight shift and the relaxation rate. If the core-polarization contribution to the Knight shift K_d is written as

$$K_d = -\frac{H_{cp}}{\beta}\chi_d,$$

where H_{cp} is called the core-polarization field, the core-polarization rate is

$$\left(\frac{1}{T_1}\right)_{cp} = 4\pi\gamma_n^2\hbar k_B T|g_d(E_F)|^2 q H_{cp}^2. \tag{8.6}$$

(The calculation[109] was done assuming cubic symmetry.) q is a numerical reduction factor (which varies between 0·2 and 0·4). The origin of this factor lies in the fact that the d-functions are orbitally degenerate and that the Knight shift involves diagonal elements of the spin operators, whereas the relaxation rate depends on the non-diagonal elements. In the relaxation rate the non-diagonal elements involving different orbital wave functions vanish and therefore the relaxation rate is reduced by a factor equal to the reciprocal of the degeneracy.

2.3. Experimental results

2.3.1. *Pure metals.* In several transition metals the Knight shift and the relaxation rate are reasonably explained by calculations based on the two-band model. In vanadium the value of the Knight shift is found to be larger than the value calculated using the Korringa relation.[110] This fact may be explained by the fact that as several contributions are present in K and $1/T_1$ the Korringa relation in the form explained in Chapter 4 has no meaning. A Korringa relation may still be derived between K_s and the contact relaxation rate and a similar relation also exists between K_d and $(1/T_1)_{cp}$, although we must take into account the presence of the reduction factor q. But for the orbital part no such relation exists because the orbital rate involves non-diagonal matrix elements of the orbital operator at the Fermi surface between d-wave functions, whereas the orbital shift involves matrix elements between the s- and d-bands for all **k** values.

A similar result is found for tungsten and molybdenum. For these metals Narath *et al.*[111,112] has tried to make a quantitative analysis of the results; for the d-electrons he uses for the density of states a value deduced from the specific-heat measurement. The density of states for the s-electrons is taken as an adjustable parameter. The d-wave functions are calculated by the tight binding method. The results for ^{95}Mo are presented in Table 8.1. Although the calculation uses only weakly justified assumptions, the results are interesting because they illustrate the possible order of magnitude of the various contributions. The three contributions to the Knight shift are found to be comparable, the

TABLE 8.1

	^{95}Mo (*Ref.* 110)		
	$(T_1 T)^{-1}$ (10^{-2} (s K)$^{-1}$)		K (%)
Contact	1·56	1·36	0·09
Core polarization	0·68	0·41	−0·11
Orbital	0·56	1·03	0·59
Total	2·8	2·8	0·57

The two figures in the column $(T_1 T)^{-1}$ correspond to the extreme values of the parameter q. The authors take as an adjustable parameter the quantity
$$\rho = g_s(E_F)/g_d(E_F).$$

	$(T_1 T)^{-1}$ (s K)$^{-1}$	
	^{105}Pd (*Refs.* 113, 109)	^{195}Pt (*Refs.* 113, 109)
Contact	0·12	17·5
Core polarization	0·15	18·1
Orbital	0·37	10·5
Total: calculated	0·64	46·1
measured	9·1	34

largest of them being the orbital contribution. The experimental results for platinum and palladium were analysed using a similar technique (they are also given in Table 8.1). For these two metals the susceptibility varies with temperature (the variation is rather large for palladium) and the Knight shifts are negative and also temperature dependent (see Fig. 8.1). The calculation of the relaxation rate predicts that the three contributions are of a comparable magnitude. As far as the absolute values are concerned the theory is in rather good agreement with the experiment for platinum, but for palladium the measured relaxation rate is an order of magnitude faster than expected.

For both these metals there is strong evidence that a simple model of a d-band of non-interacting electrons is not correct. The experimental value of the susceptibility is found to be strongly enhanced compared to the Pauli value. Moriya tried to take this effect into account, together with the peculiar shape of the Fermi surface, to explain the large observed relaxation rate (quoted in Ref. 106). If the susceptibility is strongly enhanced, it can be shown that its temperature dependence is also enhanced. This fact may be explained by the following simple calculation. Let us consider a gas of electrons in an energy band having at the Fermi energy the density of states $g_d(E_F)$. In the absence of interaction the temperature dependence of the susceptibility is given by eqn (8.5). To take the interaction into account let us use the Landau

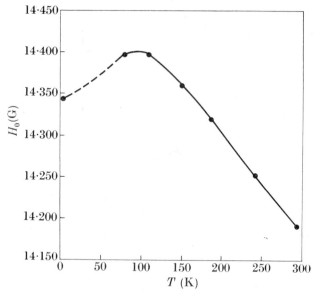

Fig. 8.1. The field for resonance at a fixed frequency ($\nu = 2\cdot683$ MHz) for Pd as a function of the temperature.[114]

model outlined in Chapter 3. In this calculation no assumption was made about the temperature (except implicitly because the Landau model loses its meaning if too many quasi-particles are present), thus eqn (3.24) remains valid, the only difference being that the quantities χ_s^F and B_0 may have a variation with the temperature. Using eqns (3.23) and (3.24) we get

$$\chi_d(T) = \frac{\chi_d^F(T)}{1+B_0(T)}\frac{m^*}{m},\tag{8.7}$$

with
$$B_0(T) = \frac{2\chi_d^F(T)}{(g\beta)^2}\langle f_e(\mathbf{k}, \mathbf{k}')\rangle,$$

or
$$\chi_d(T) = \frac{m^*}{m}\frac{\chi_d^F(T)}{1+B^0(0)\{\chi_d^F(T)/\chi_d^F(0)\}}.$$

$\chi_d^F(T)$ is the susceptibility in the absence of interactions at a temperature T. If $B^0(0)$ has a value in the vicinity of -1, eqn (8.7) predicts a very large temperature dependence (or, if one likes, the enhancement factor varies with the temperature) through the variation of the denominator. In the case of an interacting fermion gas a more fundamental calculation is possible and the result is again that the exchange interaction enhances the temperature dependence of the susceptibility.[115]

2.3.2. *The intermetallic* V_3X *compounds.* This series of compounds has been studied in great detail both experimentally and theoretically. The nuclear magnetic resonance of ^{51}V is observed and the Knight shift, the relaxation rate, and the quadrupolar splitting (the symmetry at the vanadium site is not cubic) are measured. For some of these compounds such as V_3Ga or V_3Si a superconducting state is observed when the temperature is lowered. In almost all cases the Knight shifts and susceptibilities vary with temperature, a result that is explained by assuming the presence of a very narrow d-band.[116] The relaxation rate does not vary like $1/T$, as is also expected for a very narrow band. The quadrupolar splitting is found to vary among the compounds in the same way as the density of states,[116] suggesting that a noticeable contribution to the electric field gradient is due to the electrons in the d-band.

3. Alloys of transition elements

The task of interpreting the very large number of experimental results on alloys is rather difficult as the model to use for describing a given system is never obvious; we find alloys containing very similar elements that behave differently from the point of view of magnetic properties. First we describe some cases where the results are well established.

3.1. *The copper–manganese alloys*

This alloy was the first to be studied and is certainly the alloy with which the largest number of experiments have been performed. The first results and the simplest explanation are as follows.

The behaviour of the copper nuclear resonance when a small amount of manganese is introduced is observed. The resonance line is broadened and the line width varies as the inverse of the temperature (see Fig. 8.2). The first theoretical explanation assumed that the manganese d-electrons are 'localized', they behave as the d-electron of a Mn^{++} ion in an insulating crystal. They are coupled to the conduction electrons by an exchange interaction which is written as

$$\mathscr{H} = \int d^3R \, J_{sd}(\mathbf{R}) \, \mathbf{S}_d \cdot \mathbf{S}_F(\mathbf{R}), \tag{8.8}$$

where $\mathbf{S}_d$ is the total spin of the manganese d-electrons, which are assumed to be at the origin and $\mathbf{S}_F(\mathbf{R})$ is the electronic spin density at the point R, assuming that the electrons have a plane-wave wave function. This quantity differs from the density $\mathbf{S(R)}$ defined in Chapter 3 by the absence of the function $U_{\mathbf{k}}(\mathbf{R})$ in the matrix elements (see eqn (4.22)). As these two densities are proportional the inclusion of $\mathbf{S_F(R)}$ instead of $\mathbf{S(R)}$ seems rather academic because factors like $U_{\mathbf{k}}(\mathbf{R})$ may be included in the definition of $J_{sd}(\mathbf{R})$. However, if this exchange

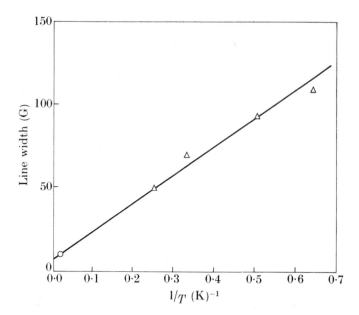

Fig. 8.2. The line width of the [63]Cu resonance in a Cu–Mn alloy (0·05 per cent Mn) as a function of the inverse temperature.[118]

integral is calculated using the d-wave function and the electronic wave functions of copper we find that the $U_{\mathbf{k}}(\mathbf{R})$ functions must be included in the calculation of $J_{sd}(\mathbf{R})$ and eqn (8.8) is established. For a more complete discussion of this point, see References 84 and 119. If, for simplicity, the range of the exchange interaction $J_{sd}(\mathbf{R})$ is assumed to be very small the interaction becomes

$$\mathscr{H} = J_{sd}\,\mathbf{S}_d\cdot\mathbf{S}_F(0), \qquad (8.9)$$

where J_{sd} is now a constant. This interaction induces an electronic magnetization varying in space and, as in the calculation of the indirect nuclear spin coupling, an indirect interaction will appear between a copper nuclear spin and the electronic spin of the manganese. The calculation is completely similar to the calculation of the Ruderman–

Kittel interaction performed in Chapter 4.[120] The following interaction is obtained:

$$\mathscr{H}_{\text{ind}} = \sum_i A(\mathbf{R}_i) \, \mathbf{S}_d . \mathbf{I}_i. \tag{8.10}$$

The function $A(\mathbf{R}_i)$ varies with R as the function $F(R)$ defined by eqn (4.50). This indirect interaction shifts the nuclear resonance line for the spin at the position $\mathbf{R}_i$. The shift is

$$\hbar\Delta\omega_i = \sum_j A(\mathbf{R}_{ij}) \langle S_{dz}^j \rangle. \tag{8.11}$$

The summation is taken over all the manganese ions. We assumed that the manganese ions are relaxing fast enough, that the spins are submitted to the thermal average field due to these ions. For a sufficient dilution the quantity $\langle S_{dz}^j \rangle$ will vary with the temperature as predicted by a Langevin equation (providing the temperature is not too low).

Equation (8.11) explains the nuclear line broadening and the temperature dependence of the line width.

The quantitative calculation of this effect is more difficult. First, let us remark that the validity of the calculation is not as good as that for the indirect nuclear spin–spin interaction. The coupling (8.9) is not as weak as the hyperfine coupling and a perturbation calculation may be a poor approximation. Secondly, the hyperfine contact coupling varies in space as a true delta function, whereas the spatial variation of $J_{sd}(R)$ was assumed to be a delta function only for simplifying the calculation. It is quite possible to take into account the spatial variation of $J_{sd}(R)$, although the calculation is more lengthy (see Ref. 84).

Using the simple theory, the value of the constant J_{sd} may be deduced from the observed experimental broadening. The value is found to be one order of magnitude larger than expected for an atomic exchange constant.

For the same alloy, Caroli and Blandin[121] gave a different treatment based on the concept of a virtual bound state.[122] Before analysing with this model the experimental results in copper–manganese alloys, let us describe this approach.

3.2. The model of the virtual bound states[122]

The problem is to study the behaviour of transitional impurities in non-transitional metals (noble metals). Because the potential $V_p(r)$ produced by the impurity is not strong enough to keep the d-electrons in real bound states they are in states called 'virtual bound states'. The energy of these states fall in the continuum of the states of the conduction electrons. For treating this situation the wave functions for the

d-electrons are described, using the phase-shift method explained in Chapter 7. A virtual bound state is characterized by a strong variation of the d-phase shift, η_2, with the energy. Such variation is shown on Fig. 8.3. More precisely, the following variation is assumed:

$$\cot \eta_2 = (E-E_0)/\Delta,$$

where E_0 is the energy of the virtual bound state and Δ its width. Looking at the wave function $F_k(r)$ of eqn (7.3) it is seen that, if $E \ll E_0$, $F_k(r)$ is a wave function for an unperturbed conduction electron, but when E is in the vicinity of E_0, $\eta_2 \simeq \frac{1}{2}\pi$, and the wave function becomes a superposition of a plane wave and a function having a d-character.

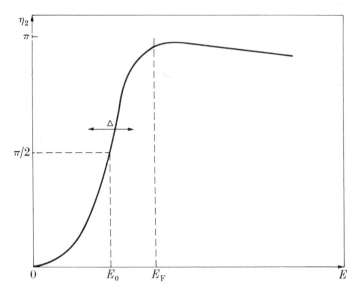

Fig. 8.3. Variation of the phase-shift η_2 as a function of the energy for a virtual bound state of energy E_0.

But in the problem we are dealing with we have to take into account the spin orientation of the d-electrons. Thus we define two phase shifts, $\eta_{2\uparrow}$ and $\eta_{2\downarrow}$, which describe the behaviour of the electrons having a given spin orientation. (If one assumes $\eta_{2\uparrow} = \eta_{2\downarrow}$ it is easy to verify that the magnetization of the impurity is equal to zero.)

If we neglect all except the d-phase shifts, they may be deduced using the Friedel sum rule

$$Z = (5/\pi)\{\eta_{2\uparrow}(E_F)+\eta_{2\downarrow}(E_F)\}, \tag{8.12}$$

and by calculating the value of the magnetic moment of the impurity. This moment is proportional to the total number of electrons pointing in one direction which, again using the sum rule, is proportional to $\eta_{2\uparrow}(E_\mathrm{F})$ minus the number pointing in the opposite direction:

$$M = \frac{5\beta}{\pi}\{\eta_{2\uparrow}(E_\mathrm{F}) - \eta_{2\downarrow}(E_\mathrm{F})\}. \tag{8.13}$$

As the two quantities Z and M are known, the phase shifts and therefore the wave functions at the Fermi level are known. Using these functions the density of spin magnetization at a given copper site is calculated, and the shift of the copper resonance line is obtained.

It is also possible, using the technique described in Chapter 7, to calculate the electric field gradient around the impurity. In the same alloy the quadrupolar splitting was estimated by measuring the intensity of the resonance line as a function of the manganese concentration.[123]

TABLE 8.2

Value of the wipe-out number n in copper–manganese alloy at 300 K. (The magnetic effect is temperature dependent and will become overwhelmingly dominant at 4 K)

	Ref.	4 MHz	8 MHz	12·8 MHz
n_exp	123	155	190	245
n_theory	121	120	161	231

As described in the previous chapter, the results are analysed by measuring the number n of copper nuclei round a manganese atom that do not contribute to the signal. Here, however, there are two effects, since the magnetic shift too may be sufficiently large to render the resonance unobservable and these spins will also contribute to n. In Table 8.2 we present the results of the measurement of n at room temperature; we notice that n increases with the magnetic field, showing clearly that a magnetic contribution is present (the second-order quadrupole splitting decreases as the field is increased). The numbers n found are well explained using the d-phase-shifted wave functions.[121]

3.3. *Alloys with non-magnetic impurities*

The behaviour discussed at some length for copper–manganese alloys is not found for all alloys with transition elements. (A similar behaviour is found, however, for copper–chromium alloys.) In other cases, like copper–nickel, aluminium–manganese, and aluminium–iron,[124,125] the results may be interpreted by assuming that there is no magnetic moment at the impurity site and only quadrupolar effects are observed. The results of the measurement of the wipe-out number are well explained using a single d-phase shift; because there is no moment, we have $\eta_{2\uparrow} = \eta_{2\downarrow} = \eta_2$.

3.4. The Kondo effect: experiments in copper–iron alloys

The problem of the Kondo effect started with a theoretical analysis of the resistivity measurements in some dilute alloys. In some cases the variation of the resistivity with the temperature presents a minimum. Kondo[126] showed that this effect is related to the scattering of the electrons by a localized magnetic impurity. This result leads to an investigation of the following basic problem: What is the structure of the ground state of a free-electron gas coupled to one localized spin by an interaction of the form given by eqn (8.8)? There are many theoretical treatments of this problem, which is not easy to handle in spite of the simplicity of the starting Hamiltonian. Some calculations lead to the following result: below a critical temperature T_K (named, of course, the Kondo temperature) the magnetization of the conduction electron cloud exactly compensates the magnetization of the impurity provided the coupling is antiferromagnetic. This state will have a very small susceptibility if the magnetic field is not too strong. We find[127]

$$\chi_S = \frac{\beta^2}{k_B T_K} \qquad \text{if } T < T_K, \tag{8.14}$$

and if the external field H_0 satisfies the inequality

$$g\beta H_0 \ll k_B T_K.$$

Let us discuss now the results of a nuclear magnetic resonance in an alloy possessing these properties. At a sufficiently high temperature, $T \gg T_K$, the results are similar to those described for copper–manganese, i.e. a large magnetic broadening whose magnitude varies as H/T is observed. But below T_K the shifts (and so the broadening) are no longer temperature dependent and their size is much reduced compared to the value given by eqn (8.11). This behaviour is indeed observed in copper–iron alloys.[128] These measurements are performed at very low temperatures, $T < 0.5$ K.

Finally, we may wonder why such an effect has not been observed in copper–manganese. A possible answer is simply that the Kondo temperature is much too low. This quantity is given approximately by the relation

$$k_B T_K \sim J_{sd} \exp\left(-\frac{1}{J_{sd} g_d(E_F)}\right). \tag{8.15}$$

J_{sd} is the exchange coupling, and hence a small decrease in the value of

J_{sd} will considerably decrease the value of the Kondo temperature. For example, the value of T_K in the system Cu–Fe is estimated to be $T_K = 10$ K, whereas in Cu–Mn we find $T_K < 0.1$ K.

3.5. *The localized spin fluctuation approach*

Let us now return to the three possible situations met in dilute alloys. We find magnetic alloys (Cu–Mn), alloys presenting a spin-compensated state that becomes magnetic if the temperature is increased, and non-magnetic alloys. The boundaries between these categories are not sharp. In some cases, like Au–V, the resistivity measurements predict a very high value for the Kondo temperature, $T_K \sim 290$ K,[129] and the nuclear magnetic resonance experiments are in agreement with this fact. We note, however, that in such a case the nuclear magnetic resonance alone is not able to discern between a non-magnetic state with a small susceptibility and a Kondo state having also a very small value for the susceptibility (when T_K is large eqn (8.14) predicts a small value for χ). It has been suggested[125,130] that these three possibilities may be described by the same model, called the localized spin fluctuation model.

Let us briefly describe the physical ideas of this model. At very low temperatures we assume that the impurity has a magnetization, but that this moment suffers very large fluctuations having a characteristic lifetime τ_0 (which does not vary with temperature). When the temperature is low, $k_B T \ll \hbar/\tau_0$, the moment fluctuates very fast, and the system behaves as a non-magnetic alloy. On the other hand, when $kT \gg \hbar/\tau_0$ the fluctuations are slower than the thermal fluctuations and the system behaves as an ordinary paramagnet. The transition occurs at a temperature $kT_K = \hbar/\tau_0$, which may be identified with the Kondo temperature.

A detailed comparison between the localized spin fluctuations approach and the spin-compensated Kondo state is not available as yet.

From the experimental point of view, the experiment performed on aluminium–manganese which we consider as an example of non-magnetic behaviour may be explained as well by using the localized spin fluctuation approach.[125,131] This alloy is specially interesting because it is not only possible to study the nuclear resonance of ^{27}Al, but also to detect the nuclear resonance signal of the impurity ^{55}Mn.

3.6. *Alloys with a non-magnetic transition metal matrix*

In all the alloys considered so far the transition element was an impurity in a non-transition metal. We shall discuss now a limited number of experiments where an impurity, which may be a transition element or another element, is embedded in a non-magnetic transition metal matrix.

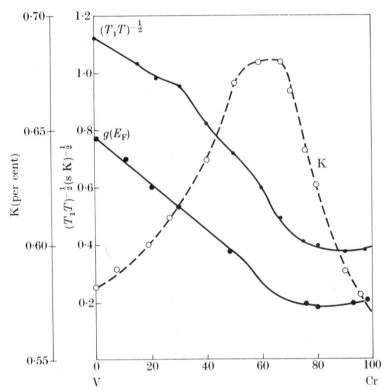

FIG. 8.4. Plot of the Knight shift and relaxation rate for ^{51}V in a V–Cr alloy. The variation of the density of states is also shown.

For some alloys involving two transition elements the results may still be interpreted using the model of a d-band described for pure metals, although we have to assume that the density of states is a function of the composition of the alloy. Such a behaviour is observed in vanadium–chromium alloys[132] where it is possible to measure the relaxation rate and the Knight shift (for ^{51}V) over almost all the range of concentrations. The relaxation rate varies exactly as the density of states (deduced again from specific heat measurements). The Knight shift presents a very different variation, a result very likely due to a large orbital contribution that is not a simple function of the density of states (Fig. 8.4). There are also many experimental results on palladium or platinum-based alloys, but detailed analysis of these problems is difficult due to their complicated band structure. It has been already noted that the susceptibility was strongly enhanced. The alloying changes this enhancement factor, in several cases reducing it (an example is vanadium in platinum, where the magnitude and the temperature dependence of the susceptibility are decreased). Interesting experiments have been[125] performed in platinum– or palladium–nickel alloys where, as there is no change in charge, only magnetic effects are observed. We observe a change in the Knight shift in the vicinity of the impurity which is due to a local change in the susceptibility, thereby locally decreasing the enhancement factor. For other alloys such as palladium–iron the enhancement factors increase and the metal becomes ferromagnetic.[107]

4. Rare-earth metals and alloys

The situation here is simplified because in almost all cases (there are some exceptions) the situation may be described using the assumption of $4f$ localized electrons.

Another new feature in the rare-earth group is that the spin-orbit coupling is large and the magnetic properties of the f-electrons are described assuming that the total angular momentum J is a good quantum number.

At low temperatures the rare-earth metals are in magnetically ordered states, the structure of which is sometimes complex. The more interesting nuclear resonance studies were performed in ordered alloys, the XAl_2 compounds, where X is a rare-earth ion. The nuclear resonance of ^{27}Al is observed. This resonance line presents a temperature-dependent shift.[133] The explanation of the occurrence of this shift has some analogy with the naïve calculation discussed for copper–manganese. An indirect interaction between the spin of the rare-earth ion and the nuclear spins is assumed. As these compounds are ordered alloys all the nuclear moments are submitted to the same local field; there is no broadening, but a shift of the resonance line. Let us assume the following interaction:

$$\mathscr{H} = \hbar \sum_{i,j} j(\mathbf{R}_{ij}) \mathbf{I}_i . \mathbf{S}_j, \qquad (8.16)$$

where $\mathbf{S}_j$ is the spin of the rare-earth ion located at the point j. This form for the interaction will be justified later. The nuclear resonance frequency is shifted by the amount $\Delta\omega$,

$$\Delta\omega = \sum_{j} j(\mathbf{R}_{ij}) \langle S_{zj} \rangle. \qquad (8.17)$$

But for a rare-earth ion, because of the large value of the spin orbit, it is convenient to express S_z as a function of J_z, using the equation

$$\langle S_z \rangle = \langle J_z \rangle \frac{\mathbf{S}.\mathbf{J}}{J(J+1)}. \tag{8.18}$$

At usual temperature only the lowest multiplicity J is occupied and the product $\mathbf{S}.\mathbf{J}$ is a negative constant for an ion with less than a half-filled shell.

The observed shifts are positive for the ions with less than a half-filled shell and negative for the others. This result in agreement with the form of the starting Hamiltonian (8.16).

Now let us discuss the value of this Hamiltonian. Such interaction will exist if an interaction between the ionic spins and the conduction electron spins of the following form exists:

$$\mathscr{H} = \sum_j \int J_{sf}(\mathbf{R}_{jk})\mathbf{S}_j.\mathbf{S}_\mathrm{F}(\mathbf{R}_k)\,\mathrm{d}^3 R_k. \tag{8.19}$$

The coupling has the same form as the coupling (8.8) and has the same origin, the exchange interaction between the localized f-electrons and the conduction electrons. This interaction, which has an electronic origin, involves the spin of the rare earth and not its magnetic moment, which would be a function $\mathbf{S}$ and $\mathbf{L}$. From the values of the observed shift an estimate of J_{sf} is deduced. Similar experiments have been performed in rare-earth phosphides.[134]

5. Magnetic metals and alloys

A large number of nuclear resonance studies were made in the three ferromagnetic metals of the iron group and in various ferromagnetic alloys. The complete understanding of the nuclear resonance properties in a magnetic metal is only possible if the properties of a ferromagnetic (or antiferromagnetic) medium are discussed in detail, and it turns out that many of these properties do not depend upon the metallic character and are equally valid in a magnetic insulator. In this book the properties of magnetically ordered states are not discussed and in this section the description will be restricted to the few properties characteristic of the metallic state.

5.1. *The hyperfine field in a ferromagnet*

In a ferromagnet of the iron group the average value of the spin operator S_z is large and well known because it is related in a very simple way to the permanent magnetic moment $g\beta S_z \simeq M$. The average hyperfine field will be large and the resonance is usually observed without applying an external field. In a sense, we are in a situation similar to that in a simple metal where the spin susceptibility χ_S is known (and therefore S_z is known). However, as we are considering transition elements, the calculation of the hyperfine field is far from being simple. There are several contributions of comparable magnitude but with opposite signs; here again a very large field is due to the polarization of the inner s-shells. The nuclear resonance technique is not the only method used for measuring the hyperfine fields; Mössbauer effect, perturbed angular correlations, and the low-temperature specific heat anomaly are also employed.

5.2. *The nuclear relaxation rates*

One of the main features of the nuclear resonance in a ferromagnet is the fact that the nuclear spins experience an oscillating field having a magnitude much larger than the magnitude of the external applied radiofrequency.[135] This enhanced field arises from the variation of the hyperfine field due to the domain wall motion or due to the rotation of the magnetization in a domain. The observed nuclear relaxation rates are different for nuclear spins located in a wall or in a domain. For spins in the domain the relaxation rate varies like $1/T$, a result that strongly suggests a modulation of the coupling by electronic motion. It is still possible to describe the system by a two-band model, but in a magnetic metal the energies (for a given wave vector) for d-electrons with up and down spins differ by a very large amount, of the order of the Fermi energy. The situation is illustrated in Fig. 8.5. (There is also a difference for the energies of the electrons in the s-band; these electrons are polarized by an exchange interaction with the d-electrons, but this energy difference is small.)

The calculation of the relaxation rate is done as for a non-magnetic transition metal, with three contributions: the contact rate, the orbital rate, and the core-polarization rate. In the calculation of this last contribution a new feature appears: the density of states for the d-electrons depends upon their spin orientations. As an example, eqn (8.6) now becomes

$$\left(\frac{1}{T_1}\right)_{\text{cp}} = 4\pi \gamma_n^2 k_B T \hbar g_{d\uparrow}(E_F) g_{d\downarrow}(E_F) q H_{\text{cp}}^2. \qquad (8.20)$$

In the case of nickel, where the up-spin band is completely filled, $g_{d\uparrow}(E_F) = 0$ and this rate vanishes.

The agreement between the calculated rate and the observed one is not very satisfactory.[136] The orbital relaxation rate is expected to be important.

5.3. *Alloys and impurities in ferromagnetic metals*

There are two possible types of studies of hyperfine fields in magnetic alloys. First, it is possible to measure the value of the field at the impurity site itself. The experiment is not usually done by nuclear resonance due to the lack of sensitivity. Secondly, by a nuclear resonance

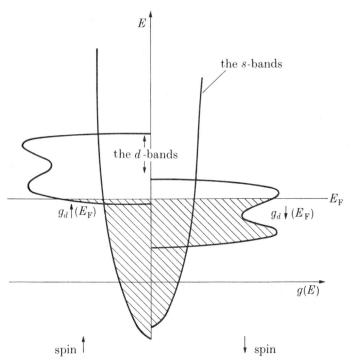

Fig. 8.5. The densities of states for a ferromagnetic metal in the two-band model.

experiment the resonance of nuclear spins near the impurity may be detected (the sensitivity is increased due to the enhancement of the driving radiofrequency). As the hyperfine field is large one expects (and finds) large shifts. But because the hyperfine field in the pure metal is difficult to calculate the theoretical interpretation of these experiments is difficult. Coming to the problem of the hyperfine field for the impurity, in some cases a simple model may be used to analyse the results. The simplest situation occurs when a non-transition metal is introduced in a ferromagnetic metal. There is a series of measurements of hyperfine fields for Ag, Cd, Sn, In, Sb, Te, and I in iron.[137] The measurements are done using the Mössbauer effect. Both positive and negative internal fields are found. (The internal fields are negative for pure iron, nickel, and cobalt.) This behaviour can be explained as follows. As we are dealing with non-transition elements the core polarization is very likely small and with a good accuracy the local field is proportional to the magnetization of the s-electrons. The conduction electrons are polarized by the exchange coupling with the magnetic d-electrons.[138] Let us call ϵ the energy required for reversing the spin orientation of the s-electron. The hyperfine field at the impurity nucleus R_i is proportional to

$$H_{\text{hfs}} = \int_0^\infty \{|\Phi_\uparrow(E, \mathbf{R}_i)|^2 \, n_\uparrow(E) - |\Phi_\downarrow(E, \mathbf{R}_i)|^2 \, n_\downarrow(E)\} \, g(E) \, \mathrm{d}E, \qquad (8.21)$$

where $\Phi_\sigma(E, \mathbf{R}_i)$ is the electronic s-wave function for a conduction electron of energy E at the point R_i for a given spin orientation. In a non-magnetic metal the wave functions for the two different spin orientations are almost equal and the field is proportional to the difference between the occupation numbers. Let us call V the strength of the potential due to the impurity that scatters the s-electrons. There are two possibilities. When V is very large compared with ϵ the potential will affect the wave function for the two spin orientations similarly, so that the difference between $\Phi_\uparrow(E)$ and $\Phi_\downarrow(E)$ may be neglected and the field is proportional to the magnetization of the conduction electrons. On the other hand, when V is small compared with ϵ the largest effect is due to the difference between the wave functions and either sign would be possible the calculation now predicts an opposite sign for the hyperfine field.

Another simple situation is the calculation of the internal field at the copper nucleus in Heussler alloys. These alloys are ordered and ferromagnetic (Cu_2MnAl, Cu_2MnSn, Cu_2MnIn) and of cubic symmetry. Large hyperfine fields are observed at the copper site. The calculation of this field is completely similar to the treatment of the shift in copper–manganese; using the d-phase shifts technique[121] the density of spin magnetization at a copper site is known. The results of the calculation are in very good agreement with the experiment.

Finally, let us discuss the temperature dependence of the local field for an impurity in a ferromagnetic alloy. In a pure ferromagnetic metal the field, being proportional to S_z, varies with the temperature in the same way as the permanent magnetic moment (although precise measurements show a slight and unexplained deviation from this behaviour).[139] In some alloys the variation of the hyperfine field at the impurity site departs drastically from the variation expected from the magnetization. This result is observed for manganese in iron. It can be explained by assuming that the spin of the manganese ion is localized

and that the hyperfine field is proportional to the average value of the spin operator $S_{z\,\mathrm{Mn}}$:
$$H_{\mathrm{Mn}} = A\langle S_{z\,\mathrm{Mn}}\rangle.$$

The manganese spin is coupled by an exchange interaction to the magnetic electrons of iron. These electrons create at the manganese site

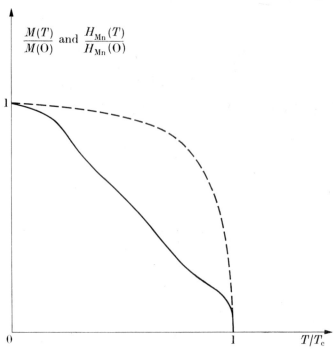

FIG. 8.6. The broken line gives the variation of $M(T)/M(0)$ for iron. The curve is the variation of the local field H_{Mn} of the manganese which obeys the law

$$H_{\mathrm{Mn}} \simeq B_{\mathrm{S}}\!\left(\frac{\beta M(T)}{k_{\mathrm{B}} T}\right)\!.$$

a field H_{ex} acting on the manganese magnetization which is proportional to the iron magnetization $M(T)$:

$$H_{\mathrm{ex}} = BM(T).$$

The average spin polarization $\langle S_{z\,\mathrm{Mn}}\rangle$ will follow the Brillouin equation:[140]

$$S_{z\,\mathrm{Mn}} = B_{\mathrm{S}}\!\left(\frac{g\beta H_{\mathrm{ex}}}{k_{\mathrm{B}} T}\right) = B_{\mathrm{S}}\!\left(\frac{g\beta BM(T)}{k_{\mathrm{B}} T}\right), \qquad (8.22)$$

and in general H_{Mn} is not simply proportional to $M(T)$. This predicted variation is in very good agreement with the measurements (see Fig. 8.6). In the same alloy the relaxation rate of $^{55}\mathrm{Mn}$ does not vary as $k_{\mathrm{B}} T$, which may also be explained by assuming a localized behaviour for the manganese spins.[141]

9

NUCLEAR RESONANCE IN
SUPERCONDUCTORS

1. Introduction to the theory of superconductors

IT is not possible in this chapter to give a detailed description of the theory of superconductors and we shall only outline the general ideas (for a complete description, see Ref. 142). There are several possible approaches for describing the properties of a superconductor. One is to start from the observed macroscopic properties and describe the system, using thermodynamic considerations together with the Maxwell equations, suitably modified. Another is to use the microscopic model proposed by Bardeen, Cooper, and Schrieffer (BCS). As the nuclear moment is a local probe, it will be more convenient to start with a short description of the microscopic BCS model. On the other hand, as a nuclear resonance experiment is performed with both static and oscillating magnetic fields, we shall have to discuss the penetration properties of these fields. The microscopic theory was illuminated by the result of a short calculation of Cooper,[143] which proved that a small attractive interaction between electrons is sufficient to produce an instability of the usual ground state of the electron gas. We shall not discuss the magnitude of this attractive interaction, which has its origin in the phonon–electron interaction.

As discussed in Chapter 3, the 'normal' electronic ground state is formed by filling all the lowest energy levels up to the Fermi energy. If we call Φ_0 the vacuum ground state, using the second quantization formalism the 'normal' ground state wave function may be written as

$$\Phi_n = \Big(\prod_{E_{\mathbf{k}\sigma} < E_F} (C^+_{\mathbf{k}\sigma}) \Big)\Phi_0. \tag{9.1}$$

The Cooper calculation shows that the electrons are grouped in pairs, the electron $\mathbf{k}\uparrow$ being attracted by the electron $-\mathbf{k}\downarrow$. Therefore in the superconducting wave function it will be assumed that if the state $\mathbf{k}\uparrow$ is occupied, the state $-\mathbf{k}\downarrow$ is also occupied. Thus we try the following type of wave function:

$$\Phi_S = \sum_{\mathbf{k}_1,\mathbf{k}_2,...,\mathbf{k}_{\frac{1}{2}N}} g_{\mathbf{k}_1} g_{\mathbf{k}_2}...g_{\mathbf{k}_{\frac{1}{2}N}}(C^+_{\mathbf{k}_1\uparrow} C^+_{-\mathbf{k}_1\downarrow} C^+_{\mathbf{k}_2\uparrow} C^+_{-\mathbf{k}_2\downarrow}...C^+_{\mathbf{k}_{\frac{1}{2}N}\uparrow} C^+_{-\mathbf{k}_{\frac{1}{2}N}\downarrow})\Phi_0, \tag{9.2}$$

where the parameters $g_{\mathbf{k}}$ are calculated by minimizing the total energy. The calculations are easier if we use a wave function of the form

$$\Phi_{\mathrm{S}} = \prod_{\mathbf{k}} (u_{\mathbf{k}} + v_{\mathbf{k}} C_{\mathbf{k}\uparrow}^{+} C_{-\mathbf{k}\downarrow}^{+}) \Phi_0. \tag{9.3}$$

This wave function differs from the wave function given by eqn (9.2) only because the total number of electrons is no longer fixed. It can be shown[142] that these two wave functions lead to the same matrix elements. Again $u_{\mathbf{k}}$ and $v_{\mathbf{k}}$ are determined by minimizing the energy of the system, taking into account the presence of an attractive potential between the electrons. To simplify the calculation BCS assumed that the potential has the following simple form. It is a constant if the electrons are in an energy shell of width $\hbar\omega_{\mathrm{D}}$ around the Fermi energy ($\hbar\omega_{\mathrm{D}} = k\Theta$, where Θ is the Debye temperature of the metal, this quantity entering because the attractive interaction arises from the phonon–electron interaction); if one or both electrons are outside this energy shell the interaction energy is assumed to be zero.

Then the following results are obtained:[142,144]

$$u_{\mathbf{k}}^2 + v_{\mathbf{k}}^2 = 1, \qquad v_{\mathbf{k}}^2 - u_{\mathbf{k}}^2 = -\frac{E_{\mathbf{k}} - E_{\mathrm{F}}}{\sqrt{\{\Delta^2 + (E_{\mathbf{k}} - E_{\mathrm{F}})^2\}}}, \tag{9.4}$$

with
$$\Delta = \frac{\hbar\omega_{\mathrm{D}}}{\sinh(1/VgE_{\mathrm{F}})} \tag{9.5}$$

if
$$|\xi_{\mathbf{k}}| \leqslant \hbar\omega_{\mathrm{D}}, \qquad \xi_{\mathbf{k}} = E_{\mathbf{k}} - E_{\mathrm{F}},$$

and
$$v_{\mathbf{k}} = 0, \qquad u_{\mathbf{k}} = 1, \qquad \text{if } \xi_{\mathbf{k}} \geqslant \hbar\omega_{\mathrm{D}},$$
$$v_{\mathbf{k}} = 1, \qquad u_{\mathbf{k}} = 0, \qquad \text{if } \xi_{\mathbf{k}} \leqslant -\hbar\omega_{\mathrm{D}}.$$

V is the value of the potential. The normal ground state may also be described by the wave function (9.3) but with the following values for $v_{\mathbf{k}}$ and $u_{\mathbf{k}}$:

$$v_{\mathbf{k}} = 0, \qquad u_{\mathbf{k}} = 1 \qquad \text{for } \xi_{\mathbf{k}} > 0,$$

which means the states above the Fermi level are empty and

$$v_{\mathbf{k}} = 1, \qquad u_{\mathbf{k}} = 0, \qquad \text{for } \xi_{\mathbf{k}} < 0,$$

i.e. the states below the Fermi level are occupied. Alternatively, the normal state can be obtained using the set of equations (9.3) and (9.4) and letting Δ tend to zero. The variations of $u_{\mathbf{k}}$ and $v_{\mathbf{k}}$ for a superconductor are shown in Fig. 9.1. We notice that the wave function Φ_{S} differs from the normal-state wave function only for electrons in the vicinity of the Fermi level. However, as many of the properties of a metal depend on the excitation of electrons in the vicinity of the Fermi level, the two states have completely different properties.

The next task is to calculate the excited states of the superconducting phase. The calculation leads to a very important result; there is now a minimum Δ for the energy of the elementary excitations, whereas in the normal state it is possible to find excitations having an arbitrary small energy.

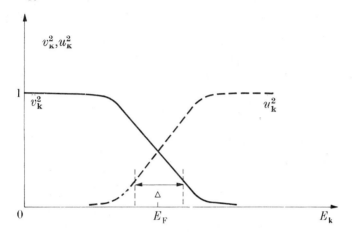

Fig. 9.1. Variation with the energy of the parameters $u_\mathbf{k}$ and $v_\mathbf{k}$ in the superconducting state.

From the knowledge of the excitations the behaviour of the superconductor at finite temperatures can be deduced. When the temperature is raised some of the excited states are occupied and if the minimization of the free energy is performed the parameters $u_\mathbf{k}$, $v_\mathbf{k}$, and Δ become functions of the temperature. The variation of Δ with the temperature is shown in Fig. 9.2. We notice that Δ decreases when the temperature increases and becomes equal to zero for a critical temperature T_c, above which the normal state is the most stable state. More details of the elementary excitations and the behaviour at finite temperatures will be given later during the discussion of the nuclear relaxation times.

Let us discuss the behaviour of the spin susceptibility in the superconducting phase. As the wave function Φ_S is built in such a way that the states with opposite spin orientations are both occupied, there is at zero degree no possibility of finding unpaired spins, and consequently $\chi_\mathrm{S} = 0$, assuming that the applied magnetic field is very small. Of course, at a finite temperature some unpaired spins are found because the excited states are occupied and the susceptibility increases with the temperature reaching the value of the normal state at $T = T_\mathrm{c}$. The variation of χ_S with temperature has been calculated by Yoshida.[145]

If the generalized susceptibility $\chi_S(\mathbf{q})$ is estimated at $T = 0$ K one finds that, for large values of q, $\chi_S(\mathbf{q})$ is the same as in the normal state, but that $\chi_S(\mathbf{q})$ tends towards χ_S, i.e. towards zero, when q becomes smaller than $1/\xi_0$.

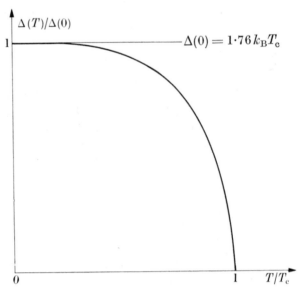

FIG. 9.2. Variation of the gap function Δ with the temperature.

The length ξ_0 is called the coherence length (this length is the radius of the Cooper-pair bound state): ξ_0 is of the order of

$$\xi_0 \simeq \frac{\hbar^2 k_F}{m^* \Delta} = \frac{1}{k_F} \frac{E_F}{\Delta}. \qquad (9.6)$$

$1/\xi_0$ is the width in wave-vector units of the region around the Fermi wave vector perturbed by the superconducting transition. For a pure metal ξ_0 is of the order of 10^4 Å. The variation of $\chi_S(q)$ is shown in Fig. 9.3.

2. The experimental aspect of a nuclear resonance experiment in a superconductor

One of the most striking properties of the superconducting state is the Meissner effect. In a superconductor a small external magnetic field does not penetrate. More precisely, the field penetrates only over a distance of the order of λ_L, where λ_L the penetration length is given by the equation

$$\lambda_L = \left(\frac{mc^2}{4\pi n e^2} \right)^{\frac{1}{2}}, \qquad (9.7)$$

where n is the number of electrons per unit volume. For a pure metal λ_L is very small, of the order of 10^{-6} cm (10^2 Å). Thus for performing a magnetic resonance experiment in a superconducting state one has to use samples whose dimensions are smaller than the penetration length.

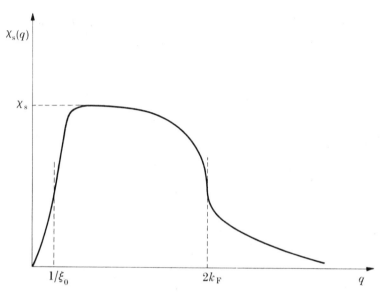

FIG. 9.3. Variation of the generalized susceptibility in a superconductor (very schematically); the dotted curve shows the susceptibility in a normal metal.

Needless to say, such conditions lead to serious experimental difficulties. Moreover, one may suspect that the properties of a sample of very small dimension (films, wires, or grains may be used) may be rather different from the properties met in bulk samples. (However, the experiments show that small samples keep almost the same properties.)

The observed macroscopic properties of a superconductor are very different, according to the relative value of the penetration length λ_L and the coherence length ξ_0. If the coherence length is the larger the superconductor is called type I, in the other case it is type II. The crucial difference between the two types is their behaviour when the external magnetic field is increased. For both types in a small magnetic field a complete Meissner effect is observed. In a type I superconductor this behaviour persists up to a critical field H_c, above which the sample becomes normal. In type II, a complete Meissner effect is observed up to a critical field H_{c_1}, but the sample becomes normal only above another larger field called H_{c_2}. Between these two fields the external magnetic field penetrates partially in the sample. Nuclear resonance experiments are very different in the two types of superconductors.

3. Knight-shift measurements in type I superconductors

The measurement of the Knight shift provides the simplest verification of the prediction of the BCS theory, as this shift (except for the orbital part) is proportional to χ_S. If the temperature is lowered below the critical temperature, the theory predicts that this shift must decrease and become zero for $T = 0$ K. In spite of the already-mentioned experimental difficulties many measurements of this quantity in type I superconductors have been reported. Experimental results are available for mercury, tin, aluminium, lead, vanadium, and so on.

Many of the measurements (more especially the first measurements) show that when the temperature is lowered below T_c, the Knight shift decreases but does not tend towards zero for $T = 0$ K. This behaviour is in contradiction with the predicted variation of χ_S. As the BCS theory is well confirmed by all the other measurements (including the variation of the nuclear-relaxation rate) one is not tempted to rule out this theory only because of this unexpected result. Two explanations have been suggested. One is to admit that in all these metals an appreciable orbital contribution to the Knight shift exists. This contribution involves all the electrons and is certainly changed only by a very small amount by the superconducting transition. This explanation may be valid for vanadium of V_3Si because the orbital contribution is known to be important when a transition element is present. On the other hand, it is difficult to believe that in all metals having a superconducting phase the orbital contribution must be large.

The other explanation is based on a modification of the BCS pairing due to the collisions of the electrons on the surface or on impurities. A theory due to Gorkov and Abrikosov[146] predicts the value of the ratio $\chi_S(T = 0)/\chi_S^n$, where χ_S^n is the susceptibility in the normal state. This ratio is a function of two parameters, the electronic mean free path and the strength of the spin-orbit coupling. The mean free path may be changed by varying either the purity of the sample or its size. There is a quantitative agreement with this modification of the BCS theory for the Knight shift and the experimental results in small particles of tin and lead.[147,148]

Finally, we note that if one finds a metal presenting a narrow electronic resonance line and a transition into a superconducting state, it will be extremely interesting to measure the susceptibility χ_S directly[149] (as explained in Chapter 4).

4. The nuclear relaxation rate in a type I superconductor

4.1. *Experimental technique*

It is rather obvious that because of the drastic change in the excitation spectrum in a superconductor, the nuclear relaxation rate (which involves excitations of the order of the nuclear Zeeman energy) will be seriously modified.

Let us first describe the experimental technique, which is rather similar to the one described in Chapter 7 for the measurement of the quadrupolar splitting in the vicinity of an impurity. The relaxation rate can be measured without observing a signal in the superconducting state, thereby avoiding the problems of penetration for the static or variable magnetic field. The amplitude of the nuclear resonance signal is measured in the normal state in a magnetic field H_0 larger than H_c (but with $T < T_c$). The field is suppressed in a very short time, short compared to the shortest relaxation time; the metal becomes a superconductor and this situation lasts a time τ. After this time τ the field is raised again to its initial value and the intensity of the resonance signal is measured. From knowledge of the variation of this intensity as a function of the time τ the relaxation rate in the superconducting state is measured.

The experiment may be performed at different temperatures, eventually at temperatures higher than T_c, for measuring the relaxation time in the normal state and in zero magnetic field. Before discussing the calculation of the relaxation rate, we note that the rate is measured in zero field and is to be compared to the rate in the normal state and in zero field. Therefore the following question may be asked: Is there a difference between the correlation coefficients in the two phases? This difference is expected to be very small. The quantity $\epsilon(\mathbf{R}_{ij})$ when $\mathbf{R}_{ij}$ corresponds to nearest neighbours depends upon the value of the susceptibility $\chi_S(\mathbf{q})$ for large values of the wave vector $\mathbf{q}$ and, as was remarked earlier, in these region $\chi_S(\mathbf{q})$ is not affected by the transition.

4.2. *Calculation of the relaxation rate*

We present here a slightly different approach from the calculation done in the BCS article.[144] We are looking for the transition probability per unit time for reversing a nuclear spin. Let us first discuss the nature and properties of the elementary excitations.

It can be shown[142] that the excitations in the superconductor are created by the operators

$$\left.\begin{array}{l} \gamma_{\mathbf{k}\uparrow}^{+} = u_{\mathbf{k}} C_{\mathbf{k}\uparrow}^{+} - v_{\mathbf{k}} C_{-\mathbf{k}\downarrow} \\ \gamma_{-\mathbf{k}\downarrow}^{+} = u_{\mathbf{k}} C_{-\mathbf{k}\downarrow}^{+} + v_{\mathbf{k}} C_{\mathbf{k}\uparrow} \end{array}\right\}. \tag{9.8}$$

$u_{\mathbf{k}}$ and $v_{\mathbf{k}}$ are defined by eqn (9.4); the destruction operators are defined as the hermitian conjugate operators: for instance,

$$\gamma_{\mathbf{k}\uparrow} = u_{\mathbf{k}} C_{\mathbf{k}\uparrow} - v_{\mathbf{k}} C_{-\mathbf{k}\downarrow}^{+}.$$

Taking into account the normalization condition $u_{\mathbf{k}}^2 + v_{\mathbf{k}}^2 = 1$, it can be shown that these operators obey the usual anticommutation rules for fermion creation and destruction operators and also that we have the relations

$$\gamma_{\mathbf{k}\uparrow} \Phi_{\mathrm{S}} = \gamma_{\mathbf{k}\downarrow} \Phi_{\mathrm{S}} = 0.$$

Therefore Φ_{S} may be considered as the vacuum wave function for these excitations. The excitation created by the operator $\gamma_{\mathbf{k}\uparrow}^{+}$ has a spin up, a wave vector $\mathbf{k}$, and an energy $\epsilon_{\mathbf{k}}$ given by the relation

$$\epsilon_{\mathbf{k}} = \sqrt{\{(E_{\mathbf{k}} - E_{\mathrm{F}})^2 + \Delta^2\}}. \tag{9.9}$$

As in Chapter 3 we define the occupation number $n(\epsilon_{\mathbf{k}}) = \gamma_{\mathbf{k}\uparrow}^{+} \gamma_{\mathbf{k}\uparrow}$, which at a temperature T is given by the equation

$$n(\epsilon_{\mathbf{k}}) = \{1 + \exp(\epsilon_{\mathbf{k}}/k_{\mathrm{B}} T)\}^{-1}. \tag{9.10}$$

The Fermi energy does not appear explicitly in the equation because the number of excitations is not fixed. This equation is more complicated than eqn (3.5), $\epsilon_{\mathbf{k}}$ being also a function of the temperature because of the variation of Δ with the temperature.

Finally, using eqn (9.9), a density of states is estimated:

$$g_{\mathrm{S}}(\epsilon_{\mathbf{k}}) = g(E_{\mathrm{F}}) \frac{\epsilon_{\mathbf{k}}}{\sqrt{(\epsilon_{\mathbf{k}}^2 - \Delta^2)}}, \tag{9.11}$$

where $g(E_{\mathrm{F}})$ is the density of states at the Fermi level in the normal state. This density is very different from the density met in a normal metal (see Fig. 9.4) and becomes infinite when $\epsilon_{\mathbf{k}} = \Delta$ (that is to say when $\mathbf{k} = k_{\mathrm{F}}$). For calculating the relaxation rate $S_{+}(\mathbf{R}_i)$ and $S_{-}(\mathbf{R}_i)$ are expressed in terms of the creation and destruction operators $C_{\mathbf{k}'\downarrow}^{+}$ and $C_{\mathbf{k}\uparrow}$ (eqn (4.22)):

$$S_{-}(\mathbf{R}_i) = \sum_{\mathbf{k},\mathbf{k}'} e^{i(\mathbf{k}-\mathbf{k}').\mathbf{R}_i} U_{\mathbf{k}}(\mathbf{R}_i) U_{\mathbf{k}'}^{*}(\mathbf{R}_i) C_{\mathbf{k}'\downarrow}^{+} C_{\mathbf{k}\uparrow}$$

which, using the transformation (9.8), becomes

$$S_{-}(\mathbf{R}_i) = \sum_{\mathbf{k},\mathbf{k}'} e^{i(\mathbf{k}-\mathbf{k}').\mathbf{R}_i} U_{\mathbf{k}}(\mathbf{R}_i) U_{\mathbf{k}'}^{*}(\mathbf{R}_i)(u_{\mathbf{k}'} u_{\mathbf{k}} \gamma_{\mathbf{k}'\downarrow}^{+} \gamma_{\mathbf{k}\uparrow} - v_{\mathbf{k}'} v_{\mathbf{k}} \gamma_{-\mathbf{k}'\uparrow} \gamma_{-\mathbf{k}\downarrow}^{+} +$$

$$+ \text{terms in } \gamma_{\mathbf{k}'\uparrow}^{+} \gamma_{-\mathbf{k}\uparrow}^{+} \text{ and } \gamma_{-\mathbf{k}'\uparrow} \gamma_{\mathbf{k}\uparrow}). \tag{9.12}$$

The products $\gamma^+_{\mathbf{k}'\uparrow}\gamma^\pm_{-\mathbf{k}\uparrow}$ or $\gamma_{-\mathbf{k}'\uparrow}\gamma_{\mathbf{k}\uparrow}$ will be neglected because they do not conserve the total energy (they involve energy changes for the electronic system which are at least 2Δ).

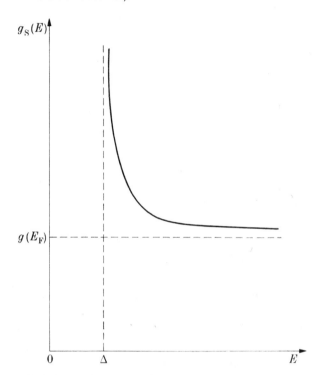

FIG. 9.4. Variation of the density of state function in a superconductor.

This equation may be written as

$$S_-(\mathbf{R}_i) = \sum_{\mathbf{k},\mathbf{k}'} e^{i(\mathbf{k}-\mathbf{k}').\mathbf{R}_i}\{u_{\mathbf{k}'}\,u_{\mathbf{k}}\,U_{\mathbf{k}}(\mathbf{R}_i)\,U^*_{\mathbf{k}'}(\mathbf{R}_i)\,\gamma^+_{\mathbf{k}'\downarrow}\,\gamma_{\mathbf{k}\uparrow}+$$
$$+v_{\mathbf{k}'}\,v_{\mathbf{k}}\,U_{-\mathbf{k}'}(\mathbf{R}_i)\,U^*_{-\mathbf{k}}(\mathbf{R}_i)\,\gamma^+_{\mathbf{k}'\downarrow}\,\gamma_{\mathbf{k}\uparrow}\} \quad (9.13)$$

or, using the relation $U_{-\mathbf{k}} = U^*_{\mathbf{k}}$, we get

$$S_-(\mathbf{R}_i) = \sum_{\mathbf{k},\mathbf{k}'} e^{i(\mathbf{k}-\mathbf{k}').\mathbf{R}_i}\{(u_{\mathbf{k}'}\,u_{\mathbf{k}}+v_{\mathbf{k}'}\,v_{\mathbf{k}})\,U_{\mathbf{k}}(\mathbf{R}_i)\,U^*_{\mathbf{k}'}(\mathbf{R}_i)\,\gamma^+_{\mathbf{k}'\uparrow}\,\gamma_{\mathbf{k}\uparrow}\}, \quad (9.14)$$

an equation quite similar to eqn (4.22), and the calculation of the rate W_{S} becomes similar to the calculation for a normal metal taking into account the new value for the energy of an excitation and of the new density of states. We obtain, neglecting constants like $\gamma_{\mathrm{n}}, g, \ldots$,

$$W_{\mathrm{S}} \simeq \iint |U_{\mathbf{k}}(\mathbf{R}_i)|^2\,|U_{\mathbf{k}'}(\mathbf{R}_i)|^2\,(u_{\mathbf{k}'}\,u_{\mathbf{k}}+v_{\mathbf{k}'}\,v_{\mathbf{k}})^2\,g_{\mathrm{S}}(\epsilon_{\mathbf{k}})\,g_{\mathrm{S}}(\epsilon_{\mathbf{k}'})\times$$
$$\times n(\epsilon_{\mathbf{k}})\{1-n(\epsilon_{\mathbf{k}'})\}\,\delta(\epsilon_{\mathbf{k}}-\epsilon_{\mathbf{k}'}-\hbar\omega_{\mathrm{n}})\,\mathrm{d}\epsilon_{\mathbf{k}}\,\mathrm{d}\epsilon_{\mathbf{k}'}. \quad (9.15)$$

The rate W_S will be compared to the rate W_n in the normal state obtained from the same equation (9.15) with $\Delta = 0$. We find

$$W_n = \iint |U_{\mathbf{k}}(R_i)|^2 \, |U_{\mathbf{k'}}(R_i)|^2 \, \{g(E_F)\} \, n(\xi_{\mathbf{k}})\{1 - n(\xi_{\mathbf{k'}})\} \times$$
$$\times \delta(\xi_{\mathbf{k}} - \xi_{\mathbf{k'}} - \hbar\omega_n) \, d\xi_{\mathbf{k}} \, d\xi_{\mathbf{k'}}.$$

The ratio of the two rates is

$$\frac{W_S}{W_n} = \frac{1}{\hbar\omega_n} \iint \frac{(\epsilon_{\mathbf{k}}\epsilon_{\mathbf{k'}} + \Delta^2)\{n(\epsilon_{\mathbf{k}}) - n(\epsilon_{\mathbf{k'}})\}}{(\epsilon_{\mathbf{k}}^2 - \Delta^2)^{\frac{1}{2}}(\epsilon_{\mathbf{k'}}^2 - \Delta^2)^{\frac{1}{2}}} \, \delta(\epsilon_{\mathbf{k}} - \epsilon_{\mathbf{k'}} - \hbar\omega_n) \, d\epsilon_{\mathbf{k}} \, d\epsilon_{\mathbf{k'}}.$$

$$(9.16)$$

In the calculation we have used the following relations:

$$n(\epsilon_{\mathbf{k}})\{1 - n(\epsilon_{\mathbf{k}} + \hbar\omega_n)\} = -kT \frac{\partial n}{\partial \epsilon},$$

$$n(\epsilon_{\mathbf{k}}) - n(\epsilon_{\mathbf{k}} + \hbar\omega_n) = -\hbar\omega_n \frac{\partial n}{\partial \epsilon},$$

and eqns (9.11) and (9.4) for $u_{\mathbf{k}}$ and $v_{\mathbf{k}}$. In eqn (9.16) there are three physical effects; the occupation numbers are changed because the energies of the elementary excitations and the density of states are different, and we note the occurrence of the factor

$$(u_{\mathbf{k}} u_{\mathbf{k'}} + v_{\mathbf{k}} v_{\mathbf{k'}})^2 = \frac{\epsilon_{\mathbf{k}}\epsilon_{\mathbf{k'}} + \Delta^2 + \xi_{\mathbf{k}}\xi_{\mathbf{k'}}}{2\,\epsilon_{\mathbf{k}}\epsilon_{\mathbf{k'}}}.$$

The term $\xi_{\mathbf{k}}\xi_{\mathbf{k'}}$ disappears because this quantity changes sign when $\mathbf{k}$ or $\mathbf{k'}$ cross the Fermi wave vector. The factor $(\epsilon_{\mathbf{k}}\epsilon_{\mathbf{k'}} + \Delta^2)/\epsilon_{\mathbf{k}}\epsilon_{\mathbf{k'}}$ varies between 1 and 2 according to the value of $\mathbf{k}$ and $\mathbf{k'}$. It may be interpreted as an interference in the superconducting state between the scattering of the electrons $\mathbf{k}$ to $\mathbf{k'}$ and of the electrons $-\mathbf{k'}$ to $-\mathbf{k}$.

Equation (9.16) may be written as

$$\frac{W_S}{W_n} = \frac{1}{\hbar\omega} \int_0^\infty \frac{\{E(E + \hbar\omega) + \Delta^2\}\{n(E) - n(E + \hbar\omega)\} \, dE}{(E^2 - \Delta^2)^{\frac{1}{2}} \{(E + \hbar\omega)^2 - \Delta^2\}^{\frac{1}{2}}}. \qquad (9.17)$$

A difficulty appears when $\hbar\omega$ tends towards zero in that the integral diverges. As the experiment is done in zero magnetic field $\hbar\omega$ is not well defined, but is of the order of magnitude of $\hbar\gamma H_L$, H_L being the local field, and this energy is very small. The divergence is due to the singularity in the density of state function. It is, however, possible to explain the observed variation of W_S over the whole temperature range $T < T_c$ by assuming that Δ is not a sharply defined quantity but that a probability $h(\Delta) \, d\Delta$ for finding Δ between the values Δ and $\Delta + d\Delta$ may be defined. For simplicity it is assumed that the function $h(\Delta)$ is constant if $\Delta_0 - \frac{1}{2}\delta < \Delta < \Delta_0 + \frac{1}{2}\delta$ and equal to zero otherwise. As the Fermi surface is not a sphere it is reasonable to assume that Δ is a function of the direction of the wave vector $\mathbf{k}$ for a given pair. In their experiments on the relaxation rate in aluminium, Hebel and Slichter[150] were able to reproduce the experimental results quite well by assuming a value of $\delta = \frac{1}{10}\Delta$. The relaxation rate when T is slightly smaller than T_c is

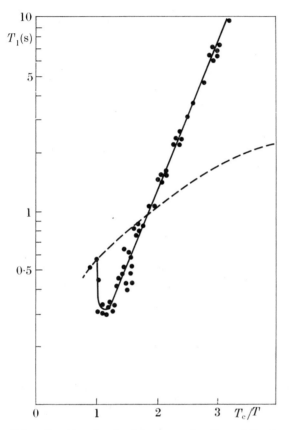

FIG. 9.5. Relaxation time in aluminium as a function of the temperature. The dotted line is the relaxation rate for a normal metal. The solid line is a calculated relaxation using a gap $2\Delta = 3 \cdot 2kT_c$, and assuming $10\delta = \Delta(T)$.[151]

greater than in the normal state, but at very low temperatures the rate decreases very fast as an exponential function of the inverse temperature (Fig. 9.5).

Another possible effect that avoids the singularity is due to the finite lifetime of the elementary excitations, which results in the density of states being no longer a singular function. It is rather difficult to decide between the two possibilities. An experiment was tried to discriminate between the two mechanisms. According to a theory due to P. W. Anderson,[152] if an anisotropy of the energy gap exists, the width δ may be reduced by adding impurities. (Roughly speaking, the impurities scatter the condensed pairs, changing the direction of their wave vectors, and only an averaged energy gap is seen.) The results of the experiment, which consist in a measurement of T_1 for pure and impure indium, do not agree very well with the predictions of the Anderson model.[152]

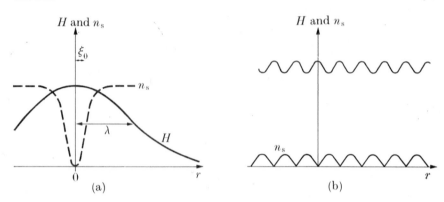

Fig. 9.6. (a) Structure of an isolated vortex line. (b) Variation of the field and the order parameter in a second-type superconductor in the vicinity of H_{c_2}.

5. Nuclear resonance in type II superconductors

5.1. *Introduction*

In a type II superconductor the penetration length is larger than the coherence length. The experiments we shall describe are performed for values of the external field H falling between the two critical fields H_{c_1} and H_{c_2} defined earlier. An experiment where H is below H_{c_1} should give the same result as in a type I superconductor. Between the critical fields the sample is divided in a two-dimensional lattice of normal state filaments parallel to the field, surrounded by superconducting regions where the currents tend to screen out the magnetic field that penetrates the normal regions. Slightly above the lower critical field H_{c_1} there exist very few normal filaments (Fig. 9.6) and the field does not penetrate very much. On the other hand, slightly below H_{c_2} the penetration is almost complete. As the distribution of the magnetic field inside the sample is not homogeneous, a broadening of the resonance line is expected. By measuring the broadening and also the shape of the nuclear resonance line some information about this vortex structure may be obtained. Another effect due to the presence of vortices is a strong modification of the excitation spectrum compared to that in a type I superconductor. This fact changes significantly the behaviour of T_1. We first discuss the line shape.[154]

5.2. The spatial dependence of the magnetic field

5.2.1. *Theory.* It is usually assumed (and confirmed by experiments) that the vortex lines are located at the corners of a regular two-dimensional lattice. All experiments show that a triangular lattice is observed

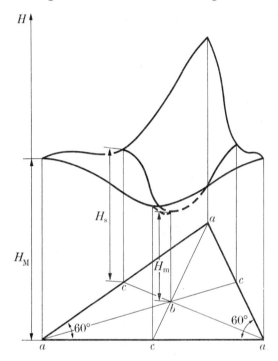

FIG. 9.7. Spatial variation of the magnetic field in a triangular vortex lattice.

in almost all cases. The variation of the field for such a lattice is shown in Fig. 9.7. A maximum field H_M is found at the vortex site a, and a minimum H_m is found at b, the centre of the triangle of three vortices. Finally, there is a saddle point at the point c half-way between two vortices (H_s). In the vicinity of H_{c_2} the spatial variation of the field and of the number n_s of superconducting electrons may be described by a set of coupled equations introduced by Landau and Ginzsburg on purely phenomenological grounds. Later on it was shown by Gorkov[155,156] that in some circumstances these equations can be justified starting from the microscopic theory. When these equations are valid and when a simple algebraic solution is found, the spatial variation of H is known and the shape of the nuclear signal may be calculated.

The amplitude of the total variation $H_M - H_m$ has the value

$$H_M - H_m = C(H - B).$$

H is the external field and B the induction defined as the spatial average value of the field in the sample.[157] C is a numerical constant which is a function of the lattice: $C = 1.46$ for a triangular lattice and $C = 1.67$

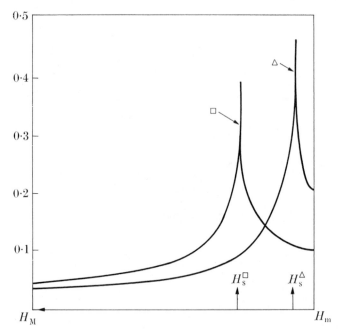

Fig. 9.8. The function $P(H)$, which gives the line shape in a type II super-conductor. The curve $\square$ is for a square lattice, and $\triangle$ for a triangular lattice. The two curves are drawn assuming the same value for $H_M - H_m$ (consequently the external fields are not the same for the two lattices) and a solution of the Ginzburg–Landau equations was used.

for a square lattice. This relation gives an estimate of the width (providing we are able to calculate B).

The function $P(H)$ that represents the number of nuclear spins submitted to a field between H and $H + dH$ is shown in Fig. 9.8. We note the singularity that appears when $H = H_s$.

The width $H_M - H_m$ has the following qualitative variation when the external field is changed. In the vicinity of H_{c_1} only a small number of vortices are present and $H_M \sim H \sim H_{c_1}$ and $H_m \sim 0$ so that the width is of the order of the lower critical field; when the field is increased the width decreases, but rather slowly, because the field H_M increases as H increases but as the number of vortices increases H_m increases also. Finally, in the vicinity of H_{c_2} the width tends to zero varying as $H_{c_2} - H$.

5.2.2. Experiments. In superconductors like V_3 Ga or NbZr, where H_{c_1} is small (and it can be shown that consequently H_{c_2} is large), the broadening remains small and can be studied by measuring the variation of the second moment of the line. Such an experiment has been done in V_3 Ga, where the agreement between the observed broadening and the theory is reasonable.[158]

Another possible way of measuring the field distribution was described by Redfield[159] and uses a cycling technique that has been described several times in this book. The nuclear resonance signal is measured in a field larger than H_{c_2} where all the sample is normal. The field is decreased to a value H lying between the critical fields. In this field the sample stays a time τ, being submitted to a radiofrequency perturbation of frequency ν. This perturbation will heat those spins that are in a local field $H(r)$ given by the equation

$$\gamma_n H(r) = 2\pi\nu.$$

Due to the vortex structure this field differs from the external field H. The external field is then increased again towards its initial value and the decrease of the signal is measured. The decrease is proportional to the number of nuclear spins submitted to the field $H(r)$, thus proportional to $P(2\pi\nu/\gamma_n)$. Keeping the same value of H the experiment is repeated using another value of ν and the whole function $P(H)$ is measured. The set of experiments is performed again using other values of H. Looking back at Fig. 9.8, we see that the position of the singularity at H_s is very different for the two types of lattice. The experimental results are in agreement with the assumption of a triangular lattice.

It is possible to measure directly the nuclear resonance signal at constant field,[160] the value of which lies between the critical fields. Because the broadening of the line varies as $H_{c_2} - H$ (in the vicinity of H_{c_2}) this technique can be applied only when H is not too far from H_{c_2}. The results are in agreement with the results obtained by the cycling method. The variation of $H_M - H_m$ in the vicinity of H_{c_2} may be calculated using the Ginzburg–Landau equations. The experimental results for this quantity are in quantitative agreement with the theory of Abrikosov using the Ginzburg–Landau equations.[161] However, very careful measurements[162] are not in agreement with this theory, the most striking fact being that H_M is not exactly equal to the external applied external field (result predicted using Ginzburg–Landau equations). This result may be qualitatively explained using a more sophisticated theory.[163]

Lastly, we mention that the line shape may also be measured using transient technique.[164]

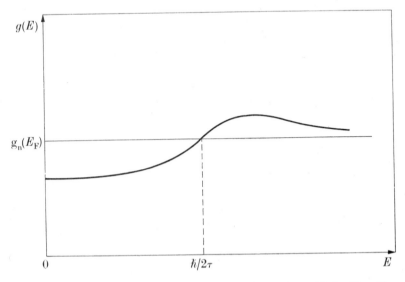

FIG. 9.9. Density of states for a dirty superconductor (type II) for $H \sim H_{c_2}$. $1/\tau$ is a damping time constant proportional to the electronic mean free path.

5.3. *The relaxation time* [153, 165, 166]

The shape of the excitation spectrum in a type II superconductor is very sensitive to the value of the electronic mean free path l. Two extreme situations are considered, dirty superconductors when $l \ll \xi_0$ and clean superconductors if $l \gg \xi_0$.

In some situations we find that there is no energy gap in the excitation spectrum. As an example, Fig. 9.9 shows the density of states for a dirty superconductor in the vicinity of H_{c_2}. This drastic change from the density of states in a BCS uniform superconductor leads to a very different behaviour for the relaxation rate. As there are not many experimental results available, we will discuss only some relatively simple cases.

5.3.1. *Superconductor with* $\kappa = \lambda/\xi_0 \gg 1$ *(type* V_3Ga*).* As λ is large the distance between vortices is always much larger than ξ_0 and in a first approximation we may consider that we have two types of nuclear moments. First there are the nuclear moments in the vortex core. These are few in number but they relax rather fast because this region behaves somewhat similarly to a normal metal and the rate is comparable to the rate $1/T_{1n}$ for a normal sample at the same temperature. The other nuclear moments relax according to the prediction of the BCS theory and at low temperature the rate is very slow. This assumption is justified by the experimental observation of a non-exponential variation of the nuclear magnetization. An exponential behaviour is again observed if the sample is rotated[167] with respect to the static field or if a large low-frequency field is added to the static field. The rotation or the modulation of the field has the effect of moving the vortex lattice with respect to a given nucleus, under which conditions the spatial average of the relaxation rate is measured. This average rate may be calculated[168] and is found to be in good agreement with the experimental results. The average rate is a function of the external field because the rate is roughly proportional to the density of vortices, which is itself a function of the applied field.

From the experimental point of view the relaxation rate may be measured using the cycling technique described for measuring the relaxation rate in type I superconductors, the only difference being that the field is not reduced to zero but to a value H between the critical fields.[169]

A spin-echo technique for measuring the recovery of a saturated resonance signal in the superconducting state may also be used.[164]

5.3.2. *Measurements in the vicinity of H_{c_2}.* Careful measurements in the vicinity of H_{c_2} clearly show a difference in behaviour between clean and dirty superconductors.[164,170] The advantage of working near H_{c_2} is that in this region a theoretical treatment is possible, using an expansion in powers of a small-order parameter. The relaxation rate is calculated in References 154, 165, and 166 for dirty and clean cases. The experimental results are in quite good agreement with the theoretical predictions.

THE SPIN RESONANCE OF CONDUCTION ELECTRONS

1. Introduction

IN this chapter the magnetic resonance experiments performed on non-localized electronic spins are described. In a metal, when nuclear moments are present nuclear magnetic resonance is always observed. The situation is completely different as far as electronic spin resonance is concerned; spin resonance of conduction electrons has been observed only in a small number of metals. This result is very likely due to the fact that in many metals the electronic-resonance line width is so large that the signal is unobservable. This chapter will begin with a comparison of the properties of electronic resonance of conduction electrons, nuclear resonance, and electronic resonance for localized electrons.

1.1. *Position of the resonance*

The electronic Zeeman energy for a free electronic spin is given by the equation

$$\hbar\omega_0 = g_s \beta H_0, \tag{10.1}$$

where β is the Bohr magneton and g_s the electronic g factor that is nearly equal to 2. This energy is about three orders of magnitude larger than a nuclear Zeeman energy for the same value of the field H_0. In an experiment in a given field the observation of the electronic resonance necessitates the use of an irradiation at a much higher frequency.

Experimentally, the values of the observed resonance frequencies for conduction electrons lead to g factors that do not differ very much from g_s. (It will be seen later that this result can be understood because, in a metal, if the factor g is very different from g_s, a broadening such that the resonance becomes unobservable is expected.) This, how- ever, is not valid for the resonance of conduction electrons in semi- conductors or semimetals, where resonances with a g factor differing considerably from the value of 2 are observed. This case will not be discussed.

1.2. *Intensity of the resonance line*

As explained in Chapter 4, the intensity of a resonance line is proportional to the static susceptibility. In a metal (see Chapter 3) this susceptibility is in order of magnitude given by the equation

$$\chi_S \simeq \frac{(g\beta)^2}{k_B T_F}.$$

This quantity will be compared with the nuclear susceptibility χ_n.

Using eqn (1.3), we get the relation

$$\chi_S = \chi_n \frac{T}{T_F} \left(\frac{g\beta}{\hbar\gamma_n}\right)^2.$$

At a temperature of 1 K with $T_F = 10\,000$ K for a typical nuclear moment the following result is obtained:

$$\chi_S = 10^2 \chi_n.$$

However, this susceptibility remains four orders of magnitude smaller than an electronic susceptibility for localized electrons at $T = 1$ K and for a comparable density of electronic spins. (This comparison is rather academic because for a comparable density an assembly of localized spins would certainly be at 1 K in a magnetically ordered state.) This increase in χ_S as compared to χ_n does not mean that the signal-to-noise ratio in a resonance experiment is increased by a factor of 100. There are several unfavourable features that tend to decrease the intensity of the electronic resonance signal. As the electronic resonance is observed using a higher frequency for the exciting field the skin depth will reduce the number of resonating electrons. If a crude comparison is made with the nuclear case, taking into account the fact that the skin depth varies as the square root of the frequency, at a fixed magnetic field thirty times fewer electronic spins than nuclear spins are seen. Another effect is that the noise factor of a spectrometer working in a microwave range is smaller than in a radiofrequency range (this reduction is typically of the order of 10). Summing up the results of this rather rough estimate, it can be seen that the signal-to-noise for the conduction-electron spin signal for the same value of the relative line width ($\Delta\omega/\omega$ or $\Delta H/H$) in a sample of large dimension is slightly less than the same quantity for nuclear spins at 1 K. If a more accurate calculation is needed it is necessary to evaluate the power absorbed by the electronic spin system and this calculation will be given in the next section.

The experimentalist may use the simple rule that the resonance line of conduction electrons can be seen only if the line width does not exceed 200 or 300 G (in a field of 3000 G).

The resonance of conduction electrons has been observed only in a limited number of metals—all the alkali metals, copper, silver, aluminium, and some divalent metals like beryllium and magnesium.

Another consequence of this discussion is that the best signal-to-noise ratio is not obtained by working in a very high external field. By using a lower frequency the skin depth is increased and the noise figure of the spectrometer improved. Indeed, in Feher and Kip's important article[171] the authors noticed that the observed signal-to-noise ratios were comparable using frequencies of 9000 and 300 MHz. Finally there is another difficulty: it is not always easy to decide whether an observed signal comes from the spins of the conduction electrons. The situation is complicated by the presence of a large number of electronic resonances having a g factor in the vicinity of 2 and a small amount of impurities may give a relatively large signal because the paramagnetic susceptibility is several orders of magnitude larger than the susceptibility in the metal. Even if there is no impurity in the sample (or in the cavity) there is still the possibility of observing resonances coming from the electronic motion (cyclotron resonances or other types of resonances). Although these resonances usually have a much larger width, their intensity is very large.

2. Calculation of the intensity and line shape

2.1. *Theory*

The first step is to discuss the properties of a resonance experiment performed on moving spins in a conducting sample. This problem was first solved by Dyson,[172] but a simpler treatment will be presented here.[173] The following simple assumption is made: the motion of the conduction electrons in a metal may be described by a diffusion equation. The variation in space and in time of the fields and magnetizations is described using the Maxwell equations.

Before writing the equations, let us note that the calculation will also be valid for the nuclear line shape if the diffusion coefficient is assumed to be zero.

The validity of a diffusion equation is not obvious. This assumption is correct if the electron suffers a large number of collisions over distances that are characteristic of our problem. It is quite possible to find experimental situations where this assumption is not valid. More details of the electronic motion will be given later in this chapter in the section devoted to the discussion of the spin wave resonances.

The electronic magnetization obeys the following equation:

$$\frac{\partial \mathbf{M}}{\partial t} = \gamma(\mathbf{H_0} + \mathbf{H_1}(t)) \wedge \mathbf{M} + D\nabla^2\mathbf{M} + \frac{1}{T_1}(\mathbf{M_0} - \mathbf{M}), \qquad (10.2)$$

where we have assumed that the time constants for the damping of the longitudinal and transverse magnetizations are equal ($T_1 = T_2$). (The generalization for $T_1 \neq T_2$ is quite straightforward.) $\mathbf{M_0} = \chi_S \mathbf{H_0}$, $\mathbf{H_0}$ is the static field, $\mathbf{H_1}$ the oscillating field, and D is the diffusion constant. We again emphasize the fact that eqn (10.2) is valid only if $\mathbf{M}$ varies slowly in space.

The Maxwell equations are

$$\mathrm{curl}\,\mathbf{H} = \frac{4\pi}{c}\mathbf{j}, \qquad \mathrm{curl}\,\mathbf{E} = \frac{1}{c}\frac{\partial}{\partial t}(\mathbf{H} + 4\pi\mathbf{M}), \qquad (10.3)$$

$$\mathbf{j} = \sigma\mathbf{E},$$

where $\mathbf{j}$ is the electric current and σ the conductivity. The last equation requires some comment. This equation is valid only if the skin depth δ is larger than the electronic mean free path Λ, δ being given by the relation

$$\delta = \frac{c}{(2\pi\sigma\omega)^{\frac{1}{2}}}. \qquad (10.4)$$

If the condition $\delta \gg \Lambda$ is not fulfilled (this situation is called the region of anomalous skin depth) it is no longer possible to write a local relation between $\mathbf{j}$ and $\mathbf{E}$ and the value of $\mathbf{j}(r)$ becomes a function of $\mathbf{E}(r')$. However the relation is still linear. In the cases we shall study, both δ and Λ remain small compared to the characteristic length of variation of $\mathbf{M}$ and, since only the relation between $\mathbf{E}$ and $\mathbf{j}$ at the surface of the sample is needed, the non-local behaviour will not change the result of the calculation. The constant of proportionality between these two quantities at the surface is equal to σ if $\delta \gg \Lambda$, but is different for other situations.

Let us consider a sample, the surface of which is infinite, with the static field perpendicular to the face (the z-axis will be taken along $\mathbf{H_0}$) and the hyperfrequency field directed along the surface. The transverse components of the magnetization M_x and M_y are assumed to remain small compared to M_z (as usual H_x and H_y are small compared to H_0). The exciting field is assumed to vary only along the z-axis, consequently the magnetization too only varies along the z-axis. The set of equations becomes

$$
\left.
\begin{aligned}
\frac{\partial M_x}{\partial t} &= -\omega_0 M_y + \gamma H_y(t) M_z + D \frac{\partial^2 M_x}{\partial z^2} - \frac{1}{T_1} M_x \\[4pt]
\frac{\partial M_y}{\partial t} &= \omega_0 M_x - \gamma H_x(t) M_z + D \frac{\partial^2 M_y}{\partial z^2} - \frac{1}{T_1} M_y \\[4pt]
\omega_0 &= \gamma H_0 \\[4pt]
j_x &= \sigma E_x, \qquad\qquad\qquad j_y = \sigma E_y \\[4pt]
\frac{\partial H_y}{\partial z} &= -\frac{4\pi}{c} j_x, \qquad\qquad \frac{\partial H_x}{\partial z} = \frac{4\pi}{c} j_y \\[4pt]
\frac{\partial E_y}{\partial z} &= \frac{1}{c}\left(\frac{\partial H_x}{\partial t} + 4\pi \frac{\partial M_x}{\partial t}\right), \quad \frac{\partial E_x}{\partial z} = -\frac{1}{c}\left(\frac{\partial H_y}{\partial t} + 4\pi \frac{\partial M_y}{\partial t}\right)
\end{aligned}
\right\} . \quad (10.5)
$$

Since we are looking for the variation of M_x, M_y, H_x, and H_y, the components of $\mathbf{j}$ and $\mathbf{E}$ are eliminated. The set of equations is simplified using the combinations $M = M_x - iM_y$ and $H = H_x - iH_y$ and looking for a solution of the following form:

$$
\left.
\begin{aligned}
M(z, t) &= M\, e^{i\omega t - kz} \\
H(z, t) &= h\, e^{i\omega t - kz}
\end{aligned}
\right\}, \quad (10.6)
$$

where M and h are two constants. The system (10.5) becomes

$$
i\omega M = i\omega_0(M - \chi_S h) + \left(Dk^2 - \frac{1}{T_1}\right) M,
$$

$$
h + 4\pi M = -\tfrac{1}{2} i\, \delta^2 k^2 h. \quad (10.7)
$$

If in the second equation M is neglected, we obtain for $H(z, t)$ the variation predicted from a skin depth effect with the skin-depth length δ. In eqns (10.7) M and h are eliminated and we obtain a second-order equation in k^2 whose roots we shall call k_1^2 and k_2^2. These two roots define two characteristic lengths; one is in the vicinity of δ and the other is of the order of $L = \sqrt{(DT_1)}$. The length L is called the diffusion length and is the distance covered by an electron during a spin lifetime.

The complete solution of the problem is written as

$$
\begin{aligned}
H(z, 0) &= h_1 e^{-k_1 z} + h_2 e^{-k_2 z}, \\
4\pi M(z, 0) &= -h_1(1 + \tfrac{1}{2} i\, k_1^2 \delta^2) e^{-k_1 z} - h_2(1 + \tfrac{1}{2} i\, k_2^2 \delta^2) e^{-k_2 z}.
\end{aligned}
\quad (10.8)
$$

k_1 and k_2 are the roots with a positive real part of the equation

$$
\left\{ i(\omega - \omega_0) + \frac{1}{T_1} - Dk^2 \right\}(1 + \tfrac{1}{2} i\, \delta^2 k^2) = 4\pi i\, \chi_S\, \omega_0. \quad (10.9)
$$

The two components h_1 and h_2 are determined by the boundary conditions. If the value of the exciting field is fixed at the surface the first condition is

$$h_1+h_2 = H(0,0) = (H_x-iH_y)_{\text{surface}}. \tag{10.10}$$

A second condition is obtained by indicating that there is no flow of magnetization across the surface.

$$\left(\frac{\partial M}{\partial z}\right)_{z=0} = 0$$

or $\qquad h_1 k_1(1+\tfrac{1}{2}i\,\delta^2 k_1^2)+h_2 k_2(1+\tfrac{1}{2}i\delta^2 k_2^2) = 0. \tag{10.11}$

The problem is now completely solved. For typical values of the parameters we find that $\delta \ll L$ and, if k_1 is the root in the vicinity of $1/\delta$ we get $h_2 \ll h_1$: only a small part of the field penetrates over the length L. The opposite situation is found for the magnetization; the larger term is $M_2\,e^{-k_2 z}$, which penetrates over a length L.

Having solved the problem of the variation of the various fields, we have to decide what quantities are measured in the experiments and consequently the experimental techniques will be discussed briefly now. There are two types of experiments.

2.2. The reflection technique

The sample is located at the end of a resonating cavity. The measured quantity is the variation of the absorbed power by the sample when the external field is varied. The power absorbed per unit of surface is given by the equation

$$P = \frac{c}{4\pi}\,\text{Re}\,(E_x H_y^* - E_y H_x^*), \tag{10.12}$$

where E_x, E_y, H_x, and H_y are the values of the fields at the surface, and P is the real part of the normal component of the Poynting vector at the surface.

If the size of the sample is very large compared to L, using eqns (10.5)–(10.11), and assuming $k_1 \gg k_2$, the following result is obtained:

$$P = P_0+\frac{\chi_S}{2}\,|h(0)|^2\omega_0\,\omega\,\delta(T_1\,T_D)^{\frac{1}{2}}\,\epsilon(x)\left\{\frac{(1+x^2)^{\frac{1}{2}}-1}{1+x^2}\right\}^{\frac{1}{2}}. \tag{10.13}$$

P_0 is the absorbed power in the absence of magnetization ($\chi_S = 0$),

$$P_0 = \frac{1}{8\pi}\,\omega\,\delta\,|h(0)|^2,$$

$$x = (\omega-\omega_0)T_1,$$

and
$$\epsilon(x) = \begin{cases} +1 & (x > 0), \\ -1 & (x < 0); \end{cases}$$

finally
$$T_{\mathrm{D}} = \frac{\delta^2}{2D}.$$

T_{D} is the time spent by an electron in the skin-depth region.
The variation of P as a function of x is shown in Fig. 10.1.

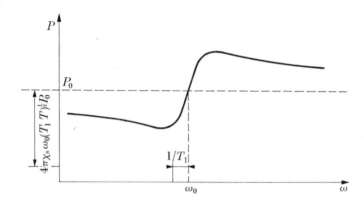

FIG. 10.1. Variation of the absorbed power as a function of the exciting frequency.[187, 188]

The result given by eqn (10.13) requires some comment. The line shape has a very peculiar form; the power absorbed by the spins is zero at resonance and the width of the line is of the order of $1/T_1$. This last result was considered at first as surprising: the spin is spending a time T_{D} in the exciting field and, using the uncertainty relation, one expects a width of the order of $1/T_{\mathrm{D}}$ (which is a very large quantity). This consideration is not correct because, due to its diffusion motion, the electron will eventually go back into the skin-depth region and suffer another excitation which has a very well-defined phase relation with the first one, and the time to be considered is not T_{D}. This experiment presents an analogy with a device due to Ramsey,[174] which is used to obtain a very narrow line in atomic-beam experiments. Another interesting quantity to consider is the value of the transverse magnetization at the surface $M(0)$, calculated when the frequency is exactly equal to the electronic resonance frequency

$$M(0) = h(0)\chi_{\mathrm{S}}\,\omega_0(T_1\,T_{\mathrm{D}})^{\frac{1}{4}}. \tag{10.14}$$

This equation may be interpreted using a time-varying susceptibility.

If we compare eqn (10.14) with the definition of the susceptibility (1.9) we get

$$\chi'' = \chi_S \omega_0 (T_1 T_D)^{\frac{1}{2}}$$

or

$$\chi'' = \chi_S \omega_0 T_1 \left(\frac{T_D}{T_1}\right)^{\frac{1}{2}}. \qquad (10.15)$$

The quantity $\chi_S \omega_0 T_1$ may be interpreted as the maximum value of the susceptibility for a system of localized spins having a resonance line width $\Delta \omega = 1/T_1$. We note that the magnetization, and therefore the signal, suffer reduction by a large factor δ/L, which is the ratio of the number of spins in the skin-depth region to the number of spins in the region where the magnetization is excited. The calculation may be extended to describe other situations, such as when the sample size is not very large compared to L. The solution will be a superposition of two decaying solutions of the type given in eqn (10.8), starting from the two faces of the sample. For this geometry the boundary conditions are that there is no flow of magnetization across the two faces and that

$$H(l, 0) = 0 \qquad \text{and} \qquad H(0, 0) = (H_x - iH_y),$$

where l is the thickness of the sample. The calculation is straightforward but lengthy.

A detailed calculation of the line shape in all the possible situations is not a purely academic problem because, as was said at the beginning of this chapter, the identification of a resonance line due to conduction electrons is sometimes difficult and if a resonance line presents exactly the shape expected, taking into account the known values of δ and L and the geometry of the experiment, such a line will be undoubtedly identified as coming from the spin of conduction electrons.

2.3. *The transmission technique*

This method takes advantage of the large difference that exists between the values of the two characteristic lengths, δ and L. The hyper-frequency power is applied at one of the faces of the sample and the induced transverse magnetization is measured at the other side.[175, 176] These two surfaces are at the ends of two resonating cavities. The thickness l of the sample is large compared to δ but small compared to L. Thus if the frequency ω is not equal to the electronic-resonance frequency, there is no coupling between the two cavities (see Fig. 10.2). But at resonance, as the magnetization is decaying only over a length L, cavity 2 is excited. In this method, instead of measuring $M(0, 0)$ we measure $M(l, 0)$; as $l < L$ these two quantities have a comparable magnitude. For experimental reasons a small testing hyperfrequency field h_2 is established in the second cavity; the quantity measured is the surface impedance of the sample in the second cavity. A complete calculation of this impedance, taking account of a possible phase difference between h_1 and h_2, is given in Reference 177.

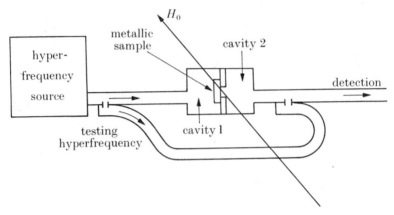

FIG. 10.2. Very schematic description of a transmission experiment.

The advantage of this method is that we can be sure that the observed signal is due to the sample, thus avoiding spurious signals due to impurities in the cavities or localized impurities in the sample. However, this method is also very sensitive to resonances due to the electronic motion.

2.4. *Influence of the external magnetic field on electronic motion*

In the previous calculation it was assumed that the electrons diffuse in an isotropic way. If the electronic mean free path is very small this assumption is valid, but for larger values of this quantity the magnetic field produces a Lorentz force on the electron and its motion is perturbed. If the electron is moving in the direction of the magnetic field there is no modification, but if it is moving at right angles the field will bend its trajectory and a smaller value for the diffusion constant is expected. This modification of the motion produces several observable effects on the transport properties. Let us call τ the average time between two collisions. The change in the diffusion coefficient will depend upon the value of the dimensionless parameter $\omega_c \tau$, where ω_c is the cyclotron resonance frequency and is of the same order of magnitude as the electronic Zeeman frequency ω_0. If $\omega_c \tau \ll 1$ the diffusion is isotropic. If this inequality is not fulfilled it will be assumed that D depends upon the relative direction of the electronic drift motion and of the magnetic field. The following equation can be justified, using a classical description of electronic motion:[178]

$$D = D_0\left\{\sin^2\Delta + \frac{\cos^2\Delta}{1+(\omega_c\tau)^2}\right\}, \tag{10.16}$$

where Δ is the angle between the two directions. This equation will be justified at the end of this chapter in the part devoted to the study of spin waves.

2.5. The line shape of nuclear resonance

For the nuclear system, D is equal to zero. The calculation using the set of equations (10.5) is easy and leads to the following result:

$$P = P_0 + \tfrac{1}{4}\chi_n|h(0)|^2\omega_n\omega\,\delta T_1\frac{1-\Delta\omega T_1}{1+(\Delta\omega T_1)^2}. \tag{10.17}$$

The line shape is an admixture of an absorption and a dispersion line shape.

However, this calculation is not valid for the nuclear line shape, because it was assumed that the transverse nuclear magnetization was varying exponentially with a time constant T_1. This assumption is equivalent to the assumption of a Lorentz shape for $\chi''(\omega)$, and in Chapter 4 we discussed situations where $\chi''(\omega)$ has a Gaussian shape. As there is no diffusion the nuclear magnetization at a given point is proportional to the oscillating magnetic field at the same point. More precisely, using the definition of the time-dependent susceptibility we obtain the relation

$$(m_x - \mathrm{i}m_y) = \{\chi_n'(\omega) - \mathrm{i}\chi_n''(\omega)\}(H_x - \mathrm{i}H_y),$$

and the set of equations (10.7) reduces to the two relations

$$m = \{\chi_n'(\omega) - \mathrm{i}\chi_n''(\omega)\}h,$$

$$h + 4\pi m = -\tfrac{1}{2}\mathrm{i}\,\delta^2 k^2 h.$$

There is only one value for k^2, because now both m and h decrease over the skin depth. Using the same boundary conditions a general expression for the power absorbed is obtained:

$$P = P_0 + \tfrac{1}{4}|h(0)|^2\omega_n\,\delta\{\chi_n''(\omega) - \chi_n'(\omega)\}. \tag{10.18}$$

3. Experimental results: g factor and susceptibility

3.1. Susceptibility

The principle of the experimental determination of susceptibility was described in Chapter 6. The experiment is difficult and as yet sufficiently accurate results are available only for lithium and sodium. Very recently this quantity was also measured for copper, but the principle of this experiment is different and will be discussed at the end of this chapter.

It is tempting to interpret the results using the Landau model (eqn (3.24)). Unfortunately there are two unknown quantities, m^* and B_0. If for m^* we use the result of the specific heat measurement, B_0 is determined. Table 10.1 gives results of such an analysis. In this chapter another method for measuring B_0 will be discussed.

TABLE 10.1

Measurement of χ_S

	Ref.	Na	Ref.	K	Li	Ref.	Cu	Ref.
χ_S/χ_S^F		1·73	42		1·73	41	1±0·1	66
		1·45	41		1·68	43		
m^*/m		1·24	179		2·19	10	1·38	66
B_0	179	(a) −0·28	42		0·29	10	0·38±0·1	66
		−0·15	41					
		(b) −0·18±0·03		−0·28±0·1				
B_1	179	0·05±0·04		−0·06±0·15				

In (a) B_0 is deduced from χ_S/χ_S^F and m^*/m.
In (b) B_0 is measured from the position of the spin-wave modes. This method gives also the value of B_1.

3.2. *The g factor*

The theoretical calculation of this quantity for a metal with a Fermi energy falling in the middle of a band is rather difficult. For alkali metals the difference between the measured g factor and g_F is small, always smaller than 10^{-2}. Except for caesium the g factors are smaller than g_F. The origin of the difference $g-g_F$ is to be found in the spin-orbit coupling. If this term is neglected the g factors will always be equal to g_F. We shall prove later that the exchange interaction between the electrons is unable to change the resonance frequency.

The spin-orbit interaction is a relativistic correction which, in the case of a free atom, takes the form

$$\mathscr{H}_{so} = \xi(r)\mathbf{l}.\mathbf{s} \tag{10.19}$$

for an electron of spin $\mathbf{s}$ and orbital moment $\mathbf{l}$. This equation is derived by assuming that the electron is moving in a potential with spherical symmetry. In a metal the spin-orbit coupling may be written as (again for one electron)

$$\mathscr{H}_{so} = \frac{\hbar}{2m^2c^2}(\mathbf{grad}\,V \wedge \mathbf{p}).\mathbf{s}, \tag{10.20}$$

where V is the total potential seen by the electron. In a metal V and therefore $\mathscr{H}_{\text{so}}$ are periodic.

Taking account of this interaction and neglecting the interactions between conduction electrons, the total one-electron Hamiltonian in the presence of an external field is

$$\mathscr{H} = \frac{1}{2m}\left(\mathbf{p}+\frac{e}{c}\mathbf{A}\right)^2 + V + \frac{\hbar}{2m^2c^2}\left\{\mathbf{grad}\,V \wedge \left(\mathbf{p}+\frac{e}{c}\mathbf{A}\right)\right\}\mathbf{s} + g_{\text{F}}\,\beta\mathbf{s}.\mathbf{H}_0,$$
(10.21)

where $$\mathbf{H}_0 = \mathbf{curl}\,\mathbf{A}.$$

The problem of finding the eigenvalues and eigenfunctions of the Hamiltonian (10.21) is very complicated, but we are not really interested in knowing all the wave functions; the problem is only to calculate the difference in energies between two levels having different spin orientations. Before describing the principle of this calculation for a metal it may be useful to recall the principle of the calculation of the g factor for a paramagnetic ion in an ionic crystal.[4]

Let us consider the following simple system. In the absence of the field the lowest state is a twofold degenerate Kramers doublet. (This situation arises for ions having an odd number of electrons.) For finding the energy levels in the presence of the magnetic field, the total Hamiltonian of the system is reduced by a canonical transformation to an effective Hamiltonian. This effective Hamiltonian describes the properties of the doublet without explicitly taking the other levels into account. (The coupling with the other levels was taken into account in the matrix S describing the canonical transformation.) Knowing this Hamiltonian, the g factor is obtained by a simple diagonalization. In a metal a similar technique is used. The Hamiltonian will be transformed into an effective Hamiltonian for each value of the wave vector $\mathbf{k}$ by a canonical transformation that involves inter-band matrix elements of the Hamiltonian (10.21) (see Ref. 180 and references therein). By this method the energy for reversing the spin of an electron of wave vector $\mathbf{k}$ in a band n is written
$$\hbar\omega_0 = (g_{\text{F}} + \delta g_{\mathbf{k}n})\beta H_0$$
with
$$\delta g_{\mathbf{k}n} = \{(X\pi_y - Y\pi_x)_{nn} + X_{nn}(\pi_y)_{nn} - Y_{nn}(\pi_x)_{nn}\}\ \text{(Ref. 181)},\quad (10.22)$$
where X is the periodic part of the coordinate operator x, or the projection on the x-axis of the periodic coordinate operator defined by eqn (4.21). The matrix elements of X are defined by

$$X_{nm} = \int_{\text{unit cell}} iU_{n\mathbf{k}}^*(r)\frac{\partial}{\partial k_x}U_{m\mathbf{k}}(r)\,d^3r.$$

$\boldsymbol{\pi}$ is the velocity operator which has the form, taking the spin-orbit coupling into account,

$$\boldsymbol{\pi} = \mathbf{p} + \frac{\hbar}{2mc^2} \mathbf{s} \wedge \operatorname{grad} \mathbf{V}.$$

It can be shown that $\delta g_{\mathbf{k}n}$ vanishes if the spin-orbit coupling is neglected. Finally, to find the observed g value, the average value of $\delta g_{\mathbf{k}n}$ for a wave vector at the Fermi surface has to be taken.

A detailed calculation has been made by Bienenstock and Brooks,[182] but the agreement with experiment is not very good. In the case of semiconductors where the electrons are in levels in the vicinity of a minimum of energy for a band, this type of calculation gives a value for the g factor in rather good agreement with the experiments. The wave functions are calculated using an expansion in powers of $\mathbf{k} - \mathbf{k_0}$, where $\mathbf{k_0}$ is the wave vector of an energy minimum. Such expansions are rather good for semiconductors. For alkali metals an expansion up to the third order in $\mathbf{k}$ is used but as δg is very sensitive to the details of the wave function such expansion may be not sufficiently accurate.

4. Line width and broadening due to impurities

4.1. *Introduction*

In all metals presenting a clearly identified spin resonance, the variation of the line width as a function of the temperature has been studied. There are two different types of behaviour: in the cases of lithium, beryllium, rubidium, and caesium the line width does not vary with the temperature; but for sodium, copper, and aluminium the line width increases when the temperature is increased. In all metals the line width is a function of the purity of the metal. The results will now be examined in detail.

4.2. *Lithium and beryllium*

The line width of the conduction-electron resonance in lithium is very narrow (the narrowest line width observed in a metal); for bulk lithium samples its value decreases if the metal is distilled until a value of the order of 0·3 G is reached.[171] Much narrower resonances are observed in small lithium particles embedded in a crystal of lithium fluoride, where these particles are produced by an intense neutron irradiation of the crystal. The lithium thus produced is presumably of very high purity. Moreover, the particles are protected from oxidation by the surrounding crystal. In these conditions, line widths as small as 0·03 G[53] are observed. For beryllium the resonance has been observed only in a sample of very high purity.

4.3. *Sodium, copper, and aluminium*

In these three metals the width varies with the temperature, the variation presenting some analogy with the variation of the resistivity in these metals. In the case of sodium[171] and copper[183] the line width at low temperatures depends upon the purity of the metal. Surprisingly, in the case of aluminium[184] the low-temperature value seems to be independent of the purity (although it has been suggested that a definite unknown impurity having a given solubility is always present in aluminium and thus the line width at low temperatures will still be due to an impurity).

4.4. *The theory of line width*

The principles of the theory of line width are described in articles by Elliot,[185] Yafet,[180] and Overhauser.[186] When a metal is placed in a magnetic field, the spin magnetization tends to line up along the field but, as in the nuclear case, this situation does not happen instantaneously. A longitudinal relaxation time T_1 is defined by an equation similar to eqn (1.12):

$$\frac{\mathrm{d}M_z}{\mathrm{d}t} = -\frac{1}{T_1}(M_z - M_0), \tag{10.23}$$

where M_0 is the equilibrium spin magnetization.

As the relaxation times T_1 are usually very short it is not convenient to use transient technique for measuring them as in the nuclear case. Usually the width of the line is measured, this width being proportional to $1/T_2$, if T_2 is the transverse relaxation time. In the case of conduction electrons there is no broadening due to the spin–spin interactions, since the motion of the electrons averages out this effect. As it is easier to calculate T_1, it is important to know how T_1 and T_2 are related. It can be shown that for an isotropic substance, if the modulation of the interaction producing the relaxation is fast, the relation $T_1 = T_2$ is obeyed. In a metal, when the Fermi surface is anisotropic we do not know the relation, but it can be shown that T_2 remains comparable to T_1 (Ref. 180, p. 69). As the calculations are far from accurate a possible difference between the two time constants will be neglected.

Let us now discuss the calculation of T_1. As in Chapter 4, the interactions between electrons will be neglected, and it will be assumed that the electrons with opposite spin orientations have different Fermi energies E_F^+ and E_F^-. If we call $w_{\mathbf{k}\uparrow,\mathbf{k}'\downarrow}$ the probability of transition from the state $\mathbf{k}\uparrow$ to the state $\mathbf{k}'\downarrow$, a quantity we have already used in Chapter 4 for calculating the nuclear relaxation rate, the rate of change in the total number of electrons with an up spin $n\uparrow = \sum_{\mathbf{k}} n_{\mathbf{k}\uparrow}$ will be

$$\frac{dn\uparrow}{dt} = \sum_{\mathbf{k},\mathbf{k}'} w_{\mathbf{k}'\downarrow,\mathbf{k}\uparrow}\, n_{\mathbf{k}'\downarrow}(1-n_{\mathbf{k}\uparrow}) - w_{\mathbf{k}\uparrow,\mathbf{k}'\downarrow}\, n_{\mathbf{k}\uparrow}(1-n_{\mathbf{k}'\downarrow}),$$

and a similar equation for $n\downarrow$. We obtain the evolution of the magnetization using eqn (3.17);

$$\frac{dM_z}{dt} = g\beta\Big\{\sum_{\mathbf{k},\mathbf{k}'} w_{\mathbf{k}\uparrow,\mathbf{k}'\downarrow}\, n_{\mathbf{k}\uparrow}(1-n_{\mathbf{k}'\downarrow}) - w_{\mathbf{k}'\downarrow,\mathbf{k}\uparrow}\, n_{\mathbf{k}'\downarrow}(1-n_{\mathbf{k}\uparrow})\Big\}. \quad (10.24)$$

The quantities $n_{\mathbf{k}\uparrow}$ and $n_{\mathbf{k}\downarrow}$ are expanded as functions of $E_{\mathrm{F}}^{+}-E_{\mathrm{F}}$ and $E_{\mathrm{F}}^{-}-E_{\mathrm{F}}$, for instance,

$$n_{\mathbf{k}\uparrow} = n_{\mathbf{k}\uparrow}^{0} + (E_{\mathrm{F}}-E_{\mathrm{F}}^{+})\frac{\partial n^{0}}{\partial E}(E_{\mathbf{k}\uparrow}),$$

where $n_{\mathbf{k}\uparrow}^{0}$ is the equilibrium occupation number. Now if we use the relation
$$M_z - M_0 = \tfrac{1}{2}g\beta \sum_{\mathbf{k}} \{(n_{\mathbf{k}\downarrow}-n_{\mathbf{k}\downarrow}^{0}) + (n_{\mathbf{k}\uparrow}^{0}-n_{\mathbf{k}\uparrow})\},$$

and again expand the change in the occupation number as a function of the change in the Fermi energies, we obtain eqn (10.23) with T_1 given by the equation

$$\frac{1}{T_1} = \Big\langle \sum_{\mathbf{k}} n_{\mathbf{k}}^{0}(w_{\mathbf{k}'\downarrow,\mathbf{k}\uparrow}+w_{\mathbf{k}'\uparrow,\mathbf{k}\downarrow}) + (1-n_{\mathbf{k}}^{0})(w_{\mathbf{k}\uparrow,\mathbf{k}'\downarrow}+w_{\mathbf{k}\downarrow,\mathbf{k}'\uparrow})\Big\rangle_{\mathbf{k}'}, \quad (10.25)$$

where the average is taken over the wave vector $\mathbf{k}'$ at the Fermi level.

Let us discuss now the various possible relaxation mechanisms. In Reference 180 four possible processes were considered. The first possibility is a relaxation rate due to the hyperfine coupling. This rate is easily estimated because the $w_{\mathbf{k},\mathbf{k}'}$ have already been calculated in Chapter 4. If the Zeeman energies are neglected eqn (10.25) becomes

$$\frac{1}{T_1} = 2\Big\langle \sum_{\mathbf{k}} w_{\mathbf{k},\mathbf{k}'}\Big\rangle_{\mathbf{k}'}, \quad (10.26)$$

or $$\frac{1}{T_1} = 2\Big\langle \sum_{\mathbf{k}} |U_{\mathbf{k}}(\mathbf{R}_i)|^2 |U_{\mathbf{k}'}(\mathbf{R}_i)|^2 \frac{2\pi}{\hbar}\Big(\frac{8\pi}{3}\gamma_{\mathrm{n}}g\beta\hbar\Big)^2 \delta(E_{\mathbf{k}}-E_{\mathbf{k}'})\Big\rangle_{\mathbf{k}'}.$$

Comparing this rate with the nuclear rate we find

$$\Big(\frac{1}{T_1}\Big)_{\mathrm{e}} \simeq \Big(\frac{1}{T_1}\Big)_{\mathrm{n}} \{k_{\mathrm{B}}\,T g(E_{\mathrm{F}})\}^{-1}.$$

This rate is temperature independent. If we consider sodium, where $T_{\mathrm{n}} = 5\,\mathrm{s}$, $T = 1\,\mathrm{K}$ assuming $1/\{k_{\mathrm{B}}g(E_{\mathrm{F}})\} \simeq T_{\mathrm{F}} \simeq 10^4\,\mathrm{K}$, one finds $(T_1)_{\mathrm{e}} = 5.10^{-4}\,\mathrm{s}$, which is a rather large value compared to the other contributions. Another mechanism is due to the interaction of an electronic spin with the current due to another electron. The Hamiltonian of this interaction may be written[186,187]

$$\mathscr{H}_{sc} = \frac{\beta e}{mc} \sum_{e,e'} \frac{1}{r_{ee'}^3} \{ \mathbf{s}_e (\mathbf{r}_{ee'} \wedge \mathbf{p}_{e'}) + (\mathbf{r}_{ee'} \wedge \mathbf{p}_{e'}) \cdot \mathbf{s}_e \}. \tag{10.27}$$

Since this interaction involves operators that are functions of two-electron variables, such terms produce processes where two electrons are scattered into two other electronic states and the relaxation process cannot be described by an equation like eqn. (10.24) but will involve probabilities $w_{\mathbf{k}\uparrow\mathbf{k}'\sigma,\mathbf{k}''\downarrow\mathbf{k}'''\sigma}$, where four electronic states have to be considered. The detailed calculations are rather complicated; we find a relaxation rate varying like $T \log T$ which, in the high-temperature range, is smaller than the rate due to the phonon scattering and usually smaller than the effects of the impurities at low temperatures.

Let us now discuss the relaxation rate due to the modulation of the spin-orbit coupling constant. The thermal vibrations have the effect of changing the periodic potential so that the electrons are no longer in a pure periodic potential, giving a probability for changing the wave vector of the electron from $\mathbf{k}$ to $\mathbf{k}'$. This mechanism is the origin of the resistivity in the high-temperature range. A change in V also involves a change in the spin-orbit coupling Hamiltonian (10.20) and consequently there is the possibility during the scattering of a change in the spin orientation. The detailed calculation is found in Reference 181. The rate due to this process varies as T at high temperatures (high compared to the Debye temperature) and as T^5 at low temperatures. It can be shown that this rate involves matrix elements in $w_{\mathbf{k},\mathbf{k}'}$ having the same form as the matrix elements involved in the calculation of the resistivity and indeed the same temperature dependence is found. The agreement between the result of this calculation and the experimental width for sodium is not very good. In all the metals presenting a temperature dependence of the line width (Na, Cu, and Al) we find that this width varies as the resistivity.

Finally, the line width may be due to a scattering by the impurities. This process will be further discussed in the next section.

It was also suggested that the scattering at the surface may change the spin orientation. This assumption was experimentally tested in copper[184] where a definite observation was made of the effect of the surface scattering. In the experiments using sodium or lithium, however, there is no clear evidence of a broadening due to the surface scattering.

We have already noted that for lithium the narrowest lines are observed for rather small particles.

Lastly, it is important to note that although the two scattering mechanisms, phonons and impurities, both contribute to the resistivity and to the line width, there is no reason to expect that these two quantities are proportional in the whole temperature range. To calculate the phonon contribution to the line width we consider the modulation of the spin-orbit coupling for the pure metal, whereas for the impurity scattering the spin-orbit energy could arise from the impurity itself.

The problem of impurity scattering may be studied experimentally by measuring the variation of the line width produced by adding known quantities of impurities. It is these experiments that we shall discuss now.

4.5. *Experimental study of impurity scattering*

Systematic studies of impurity scattering have been made using the variation of the resonance line width for sodium, lithium,[188,189] and copper[190] as a function of the concentration of impurities. A linear variation of the line width as a function of the impurity concentration is observed. From the measurement of the slope of this curve a cross-section for the spin-flip scattering is derived. Let us again neglect the interaction between conduction electrons. The total probability for flipping the spin of an electron at the Fermi level is given by the equation

$$W_{\uparrow\downarrow} = \left\langle \sum_{\mathbf{k}'} w_{\mathbf{k}\uparrow,\mathbf{k}'\downarrow} \right\rangle_{\mathbf{k}}. \tag{10.28}$$

The mean free path $l_{\uparrow}$ for the spin of the electron is related to this probability by

$$v_{\mathrm{F}}/l_{\uparrow} = W_{\uparrow\downarrow}, \tag{10.29}$$

where v_{F} is the velocity for electrons having the Fermi energy. The cross-section $\sigma_{\uparrow\downarrow}$ may be calculated as a function of $l_{\uparrow}$. If there are $N_0 c$ scattering centres per unit volume (c is the impurity concentration and N_0 the number of atoms per unit volume), we write that around each centre the volume $l_{\uparrow}\sigma_{\uparrow\downarrow}$ is covered by the spin between two collisions; thus $N_0 c l_{\uparrow}\sigma_{\uparrow\downarrow}$ must be equal to the unit volume and

$$\sigma_{\uparrow\downarrow} = \frac{1}{N_0 c l_{\uparrow}}. \tag{10.30}$$

For elastic scattering by the impurity it may be found that

$$w_{\mathbf{k}'\downarrow,\mathbf{k}\uparrow} = w_{\mathbf{k}\uparrow,\mathbf{k}'\downarrow}.$$

This relation may be derived using eqn (10.24) and writing that at equilibrium $\mathrm{d}M_z/\mathrm{d}t = 0$; thus

$$\frac{w_{\mathbf{k}\uparrow,\mathbf{k}'\downarrow}}{w_{\mathbf{k}'\downarrow,\mathbf{k}\uparrow}} = \frac{1-n^0_{\mathbf{k}\uparrow}}{n^0_{\mathbf{k}\uparrow}}\frac{n^0_{\mathbf{k}'\downarrow}}{1-n^0_{\mathbf{k}'\uparrow}} = \exp\{(E_{\mathbf{k}\uparrow} - E_{\mathbf{k}'\downarrow})/k_{\mathrm{B}}T\}.$$

This ratio is equal to unity for an elastic collision. Using (10.25) we obtain

$$\frac{1}{T_1} = W_{\uparrow\downarrow} + W_{\downarrow\uparrow},$$

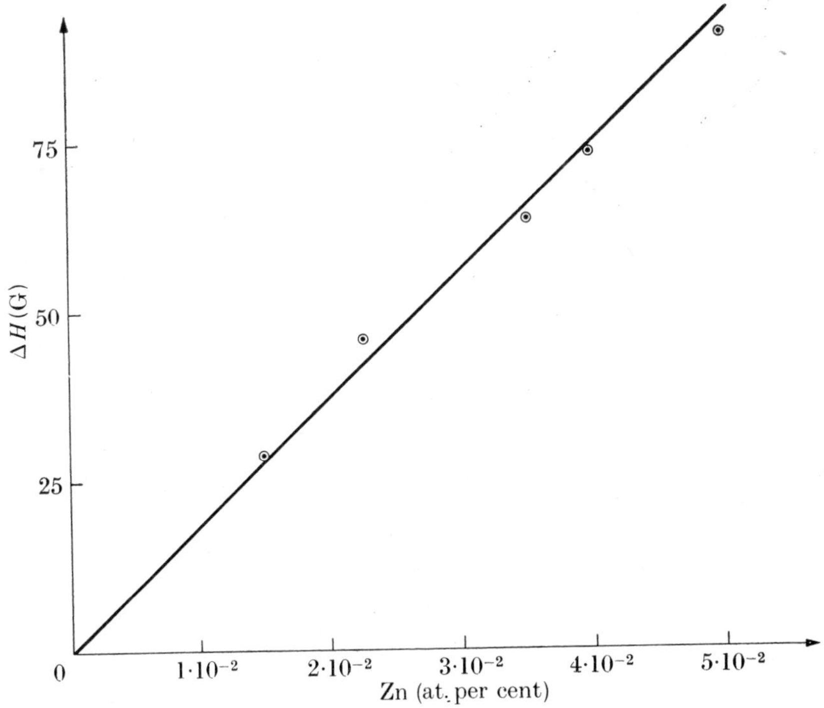

FIG. 10.3. Line width of the resonance line of Li in Li–Zn alloys as a function of the Zn concentration.

where $W_{\uparrow\downarrow}$ is defined by (10.28) and $W_{\downarrow\uparrow}$ similarly. The cross-section may be expressed as a function of $1/T_1$ (assuming $W_{\uparrow\downarrow} = W_{\downarrow\uparrow}$),

$$\sigma_{\uparrow\downarrow} = \frac{1}{2N_0\,v_{\mathrm{F}}}\frac{\partial}{\partial c}\left(\frac{1}{T_1}\right). \tag{10.31}$$

(In Refs. 188 and 189 the authors use another definition for the spin-flip scattering $\sigma_{\mathrm{sf}} = \sigma_{\uparrow\downarrow} + \sigma_{\downarrow\uparrow} \simeq 2\sigma_{\uparrow\downarrow}$.) Systematic measurements of σ_{sf} were made in lithium and the results (see Fig. 10.3) are analysed as follows. Let us first consider the situation where there is no difference between the valency of the metal and the impurity. The wave functions of electrons at the Fermi level are built by orthogonalizing a plane wave function with core wave functions of the impurity:

$$|\mathbf{k}_F, \sigma\rangle = |e^{i\mathbf{k}_F \cdot \mathbf{r}} - \sum_c a_c \phi_c, \sigma\rangle, \qquad (10.32)$$

where ϕ_c is an atomic wave function of the impurity and a_c is an overlap integral, $a_c = \langle \phi_c | e^{i\mathbf{k}_F \cdot \mathbf{r}} \rangle$. The probability of reversing the spin is calculated using the spin-orbit interaction of the impurity as a perturbation. $\langle \mathbf{k}_F, \uparrow | \mathscr{H}_{so} | \mathbf{k}'_F, \downarrow \rangle$ is calculated considering only the influence of the atomic part of the wave function (10.32) and the agreement for the cases of LiAg, LiAu, and NaAu is found to be very good.

When there is a difference in valency, the screening effect discussed in Chapter 6 has to be taken into account. Asik, Ball, and Slichter[188] replace the impurity by a screened Coulomb potential acting on the conduction electrons, and the screening length is adjusted in such a manner that the phase shifts obey the Friedel sum rule.

This phase-shifted electronic wave function is again orthogonalized to the core wave function (the largest contribution is due to the overlap with the p wave functions). Finally, the matrix elements of the spin-orbit coupling are evaluated and thus also the spin-flip cross-section.

The agreement is rather good if the valency differences are small ($\Delta Z = \pm 1$), but for larger values of ΔZ ($\Delta Z = 3$) large discrepancies are found. A much better agreement is obtained if the assumption of the presence of a virtual p bound state is made.[191] In Chapter 8 the concept of virtual bound states was defined. Here only the p phase shift is considered. The scattered wave function is described by considering the phase shifts for the total angular momentum j. For a p wave two phase shifts are introduced $\eta_1^{\frac{3}{2}}$ and $\eta_1^{\frac{1}{2}}$. The spin-flip cross-section is given by the equation of Mott and Massey,[192]

$$\sigma_{\uparrow\downarrow} = \frac{8\pi}{3k_F^2} \sum_l \sin^2(\eta_l^{l+\frac{1}{2}} - \eta_l^{l-\frac{1}{2}}), \qquad (10.33)$$

which becomes, for p electrons,

$$\sigma_{\uparrow\downarrow} = \frac{16\pi}{9k_F^2} \sin^2(\eta_1^{\frac{3}{2}} - \eta_1^{\frac{1}{2}}). \qquad (10.34)$$

The difference between the phase shifts is estimated, using the equation for the phase shift of a virtual bound state,

$$\cot \eta_1^j = \frac{\epsilon - \epsilon_j}{\Delta},$$

and the value of the difference $\epsilon_{\frac{3}{2}} - \epsilon_{\frac{1}{2}}$, which is proportional to the spin-orbit constant λ_p,

$$\lambda_p = \tfrac{2}{3}(\epsilon_{\frac{3}{2}} - \epsilon_{\frac{1}{2}}).$$

Using these equations the cross-section is found to be

$$\sigma_{\uparrow\downarrow} = \frac{4\pi}{k_F^2}\left(\frac{\lambda_p}{\Delta}\right)^2 \sin^4\eta_1, \qquad 2\eta_1 = \eta_{\frac{3}{2}}^1 + \eta_{\frac{1}{2}}^1.$$

Ferrell and Prange[192] estimated η_1 using the Friedel sum rule and found a very good agreement for all the alloys, using the value of $\Delta = 3$ eV.

The experimental results for copper[190] alloyed with non-transitional elements such as zinc, gallium, germanium, or arsenic were analysed, using the same method, and a reasonable agreement was also obtained.

For transitional impurities in copper the results are analysed using the d-phase shifts, $\eta_{\frac{3}{2}}^1$ and $\eta_{\frac{5}{2}}^1$; but, as discussed in Chapter 8, one must take into account the difference between the diffusion by d-states with spin up or down. The cross-section is a function of four d-phase shifts. These four quantities are calculated using the Friedel sum rule, the values of the magnetization (for a magnetic impurity), the spin-orbit constant, and the width of the virtual level. Good agreement is obtained with the observed spin-flip cross-section as with the value of the excess resistivity. Finally, we shall describe an indirect method for measuring the spin-flip cross-section which does not necessitate the observation of the conduction electron resonance line.[193] We consider the nuclear resonance of copper in copper–manganese alloys. As explained in Chapter 8, we know how to calculate the oscillations of the electronic magnetization and thus to predict the value of the copper nuclear line width. If another non-magnetic impurity is added, the mean free path for the spin of the conduction electrons will be shortened, effecting a change in the value of the variation of the magnetization, and a change of the nuclear line width will be observed. From the value of this change an estimate of the spin-flip cross-section may be deduced.

5. Spin wave resonances

By studying very carefully the shape of the resonance line in sodium, using the transmission technique, Schultz[194,195] noticed the occurrence of a series of satellite lines around the main peak. These resonances are interpreted using the theory of the elementary excitations of an interacting electron gas. The satellite peaks correspond to the excitation of non-uniform modes ($q \neq 0$), whereas the main peak is a spatially uniform excitation of the spin magnetization.

The calculation of this effect made by Platzmann and Wolff,[179] uses an extension of the Landau model due to Silin.[196,197] First the Landau theory given in Chapter 3 must be generalized for describing slightly non-uniform situations.

In Chapter 3 the quantities $\delta n_{\mathbf{k}\sigma}$ were defined (for the moment the spin index will be omitted). Let us consider a situation where the quantity $\delta n_{\mathbf{k}}$ is a slowly spatially varying function and call its spatial Fourier transform $\delta n_{\mathbf{k}}^{\mathbf{q}}$. A quantity like $\delta n_{\mathbf{k}}^{\mathbf{q}}$ is only meaningful for values of $\mathbf{q}$ much smaller than the Fermi wave vector $\mathbf{k}_{\mathrm{F}}$. A transport equation for $\delta n_{\mathbf{k}}(\mathbf{r})$ will be established. The energy of a quasi-particle becomes a function of $\mathbf{r}$ and eqn (3.11) is now written

$$\epsilon(\mathbf{k},\mathbf{r}) = \epsilon_0 + \sum_{\mathbf{k}'} f(\mathbf{k},\mathbf{k}')\,\delta n_{\mathbf{k}'}(\mathbf{r}). \qquad (10.35)$$

This equation is valid provided the range of the interaction is very short, an approximation certainly very good for liquid ^{3}He but very questionable for electrons, due to the long range of the Coulomb interaction. However, the corrections due to the long range of the interaction were calculated by Silin[196,197] and found to have no effect on the spin excitations (though they have very large effects on the behaviour of charge fluctuations).

We shall consider the quasi-particle as a classical particle having the classical Hamiltonian (10.35). This particle has a velocity $v_{\mathbf{k},\alpha}$ given by the relation

$$v_{\mathbf{k},\alpha} = \frac{1}{\hbar}\frac{\partial\epsilon}{\partial k_\alpha} \quad (\alpha = x,y,z). \qquad (10.36)$$

As the energy is a function of $\mathbf{r}$, the quasi-particle will diffuse towards regions where the energy is minimum, an effect that may be described by a force $\mathbf{F}$ acting on the quasi-particle. This force is given by the equation

$$\mathbf{F} = -\mathbf{grad}\,\epsilon(\mathbf{k},\mathbf{r}). \qquad (10.37)$$

In the absence of collisions the number of quasi-particles per unit volume in the six-dimensional $\mathbf{k},\mathbf{r}$ space is a constant and the classical transport equation is thus obtained:

$$\frac{\partial}{\partial t}n_{\mathbf{k}}(\mathbf{r},t) + \frac{1}{\hbar}\sum_\alpha\left(\frac{\partial n}{\partial r_\alpha}\frac{\partial\epsilon}{\partial k_\alpha} - \frac{\partial n}{\partial k_\alpha}\frac{\partial\epsilon}{\partial r_\alpha}\right) = 0. \qquad (10.38)$$

We are only interested in small variations of n around the equilibrium value $n_{\mathbf{k}}^0$ and we will define the quantity $\delta n_{\mathbf{k}}(\mathbf{r},t)$ as

$$\delta n_{\mathbf{k}}(\mathbf{r},t) = n_{\mathbf{k}}(\mathbf{r},t) - n_{\mathbf{k}}^0.$$

Equation (10.38) becomes, if only terms linear in $\delta n_{\mathbf{k}}$ are considered,

$$\frac{\partial}{\partial t} \delta n_{\mathbf{k}}(\mathbf{r}, t) + \sum_\alpha v_{\mathbf{k}\alpha} \frac{\partial}{\partial r_\alpha} \{\delta n_{\mathbf{k}}(\mathbf{r}, t)\} +$$

$$+ \sum_\alpha \left\{ v_{\mathbf{k}\alpha} \delta(\epsilon_{\mathbf{k}} - E_{\mathrm{F}}) \sum_{\mathbf{k}'} f(\mathbf{k}, \mathbf{k}') \frac{\partial}{\partial r_\alpha} \delta n_{\mathbf{k}'}(\mathbf{r}, t) \right\} = 0. \quad (10.39)$$

The delta function arises from the term $\partial n^0/\partial k_\alpha$, which may be written

$$\frac{\partial n^0}{\partial k_\alpha} = \frac{\partial n^0}{\partial \epsilon_{\mathbf{k}}} \frac{\partial \epsilon_{\mathbf{k}}}{\partial k_\alpha} = -\hbar v_{\mathbf{k}\alpha} \delta(\epsilon_{\mathbf{k}} - E_{\mathrm{F}})$$

at low temperature.

In Chapter 3 the spin variable was introduced by defining the two quantities, $n_{\mathbf{k}\uparrow}$ and $n_{\mathbf{k}\downarrow}$. These quantities are sufficient to describe situations without an external magnetic field, because in this condition all directions are equivalent. In the presence of a field, one has to consider not only the component of the magnetization along the field, but also the transverse components that involve non-diagonal elements with respect to the spin variables. We must therefore describe our quasi-particles $\mathbf{k}$ by a 2×2 matrix, which is an operator in the spin space. This matrix expressed in terms of the Pauli matrices will be written as

$$\rho_{\alpha\alpha'}(\mathbf{k}, \mathbf{r}) = n_{\mathbf{k}}(\mathbf{r}) I_{\alpha\alpha'} + \mathbf{M}(k, r) . \boldsymbol{\sigma}_{\alpha\alpha'},$$

where I is a unit matrix, $I_{\alpha\alpha'} = \delta_{\alpha\alpha'}$, and σ_x, σ_y, σ_z are the Pauli matrices. If we consider the diagonal term, we have the two relations

$$n_{\mathbf{k}}(r) + M_z(\mathbf{k}, \mathbf{r}) = n_{\mathbf{k}\uparrow}(\mathbf{r}),$$

$$n_{\mathbf{k}}(r) - M_z(\mathbf{k}, \mathbf{r}) = n_{\mathbf{k}\downarrow}(\mathbf{r}).$$

We shall study the motion of the transverse components of the vector $\mathbf{M}$ or, more precisely, of the quantity

$$M_+(\mathbf{k}, \mathbf{r}) = M_x(\mathbf{k}, \mathbf{r}) + iM_y(\mathbf{k}, \mathbf{r}).$$

The quasi-particle Hamiltonian now becomes an operator for the spin variables. Silin[196,197] establishes the transport equation for the matrix ρ, from which the evolution of M_+ is deduced using the relation

$$M_+ = \tfrac{1}{2}\mathrm{Tr}\{(\sigma_x + i\sigma_y)\rho\}.$$

(The trace is taken over the spin variables.) Instead of studying M_+, Wolff and Platzmann[179] introduced a function $g_{\mathbf{k}}(r)$ defined by the relation

$$M_+(\mathbf{k}, \mathbf{r}) = -\frac{\partial n_0}{\partial \epsilon_{\mathbf{k}}} g_{\mathbf{k}}(\mathbf{r}) = \delta(\epsilon_{\mathbf{k}} - E_{\mathrm{F}}) g_{\mathbf{k}}(\mathbf{r}).$$

This function has the advantage that the quasi-particles that enter the equation are automatically at the Fermi level.

Finally, in the transport equation the Lorentz force acting on the quasi-particle must be added (eqn (10.37)). The following transport equation is obtained (see Ref. 10):

$$\frac{\partial}{\partial t}g_{\mathbf{k}} + \sum_{\alpha}\left\{v_{\mathbf{k}\alpha}\frac{\partial}{\partial r_{\alpha}} + \frac{e}{c}(\mathbf{v}\wedge\mathbf{H}_0)\frac{\partial}{\partial k_{\alpha}} + i\Omega_0\right\}(g_{\mathbf{k}}+\delta g_{\mathbf{k}})$$
$$= \text{damping terms} + \text{excitation terms.} \quad (10.40)$$

In this equation we have defined Ω_0 to be the difference in energy of two quasi-particles $\mathbf{k}$ with opposite spin orientations. As discussed in Chapter 3 in the calculation of the susceptibility, Ω_0 differs from ω_0 because to the external field must be added the exchange field due to the other quasiparticles. Ω_0 has the value

$$\Omega_0 = \frac{\omega_0}{1+B_0}; \quad (10.41)$$

$(e/c)\mathbf{v}\wedge\mathbf{H}_0$ is the Lorentz force; $\delta g_{\mathbf{k}}$ is a term that is due to the presence of the exchange interaction in the Hamiltonian

$$\delta g_{\mathbf{k}} = \tfrac{1}{2}\sum_{\mathbf{k}'}f_e(\mathbf{k},\mathbf{k}')\,g_{\mathbf{k}'}\,\delta(\epsilon_{\mathbf{k}'}-E_{\mathrm{F}}). \quad (10.42)$$

The factor $\tfrac{1}{2}$ in front of the right-hand side of this equation is due to our definition of f_e using eqn (3.12). Pines and Nozières[10] use a quantity f_a which is twice f_e. Generally speaking, different authors use different definitions and much care has to be exercised to obtain the correct numerical factors.

The excitation term that describes the force due to the hyperfrequency field has the form[179]

$$\tfrac{1}{2}\gamma_e\left(i\Omega_0 + \sum_{\alpha}v_{\mathbf{k}\alpha}\frac{\partial}{\partial r_{\alpha}}\right)h_+(\mathbf{r},t).$$

Now the structure of the damping term has to be discussed. As the electronic line for uniform excitation is very narrow the spin lifetime for a uniform excitation is very long and in first approximation the damping may be neglected. But the collision will change the electronic space variation and therefore affect $g_{\mathbf{k}}(r)$ when this quantity is not uniform. As the average magnetization is not affected by the collision, the simplest possibility for a damping term is

$$-\frac{1}{\tau}\{g_{\mathbf{k}}(\mathbf{r})-\langle g_{\mathbf{k}}(\mathbf{r})\rangle\},$$

where $\langle g_{\mathbf{k}}(\mathbf{r})\rangle$ is the spatial average of $g_{\mathbf{k}}(\mathbf{r})$ and τ is a time of the order of the average time between two collisions. Equation (10.40) may be used to calculate the signal in a transmission experiment.

The results are more conveniently analysed using the generalized susceptibility $\chi_S(\mathbf{q}, \omega)$.

The calculation of the transmitted signal S_T will be performed in two steps. First, S_T will be expressed as a function of $\chi_S(\mathbf{q}, \omega)$. Using the transport equation $\chi_S(\mathbf{q}, \omega)$ will be estimated. In a transmission experiment when the skin depth δ_s is very small the exciting field may be written as

$$h_+(\mathbf{r}, t) = h\,\delta_s\,e^{i\omega t}\delta(z),$$

where the z-axis is perpendicular to the plate and the origin is in the face of the sample in the first cavity. The spatial Fourier transform of this function has a very simple form:

$$h_\mathbf{q}(t) \simeq h\,\delta(q_x)\,\delta(q_y)\,e^{i\omega t}.$$

The observed signal is proportional to the value of the magnetization in the plane $z = l$,

$$S_T = \sum_{\mathbf{k}, q_z} M^\mathbf{q}_{+\mathbf{k}}(q_x = q_y = 0)\,e^{iq_z l}, \tag{10.43}$$

where $M^\mathbf{q}_{+\mathbf{k}}$ is the Fourier transform of $M_{+\mathbf{k}}(\mathbf{r})$. Taking into account the value of $h_\mathbf{q}$ and the definition of $\chi_S(\mathbf{q}, \omega)$, the signal is written as

$$S_T \simeq \sum_{q_z} \chi_S(q_z, q_x = q_y = 0, \omega)\,e^{iq_z l}. \tag{10.44}$$

As l is a macroscopic length, $\chi_S(\mathbf{q}, \omega)$ has to be calculated for small values of $\mathbf{q}$. The calculation of $\chi_S(\mathbf{q}, \omega)$ using eqn (10.40) is long and only the principle will be discussed. Equation (10.40) is written, using $g^\mathbf{q}_\mathbf{k}(\omega)$, the Fourier transform in space and time of $g_\mathbf{k}(\mathbf{r}, t)$. For a given value of $\mathbf{q}$ we have a system of coupled linear equations corresponding to the various values of the vector $\mathbf{k}$ (its magnitude is equal to k_F). Neglecting the excitation and damping terms, we look for the eigenvalues ω of this system. Two types of solutions are found corresponding to different excitations.

First, for some excitations one of the $g^\mathbf{q}_\mathbf{k}$ has a large amplitude, whereas all the others have vanishingly small values. (In this case, in the equations for $g^\mathbf{q}_\mathbf{k}$ the term $\delta g^\mathbf{q}_\mathbf{k}$ may be neglected.) The energies of these excitations are

$$\omega_{\mathbf{k}, \mathbf{q}} = \Omega_0 + v_F q \cos\theta_{\mathbf{k}, \mathbf{q}},$$

where $\theta_{\mathbf{k}, \mathbf{q}}$ is the angle between the vectors $\mathbf{q}$ and $\mathbf{k}$. (The Lorentz force has been neglected; for a more detailed treatment see References 179, 195.) This energy may be simply interpreted as the difference between $\epsilon_{\mathbf{k}+\mathbf{q}\uparrow}/\hbar$ and $\epsilon_{\mathbf{k}\downarrow}/\hbar$, the excitation being that of an electron of wave vector $\mathbf{k}$ which has its momentum increased by $\mathbf{q}$ and its spin direction changed.

Another type of excitation is found when all the $g_{\mathbf{k}}^{\mathbf{q}}$ have comparable magnitudes. We shall discuss now this collective excitation. For small values of $\mathbf{q}$, (10.40) becomes

$$-i\omega g_{\mathbf{k}}^{\mathbf{q}}+i(v_{\mathbf{k}}\, q\cos\theta_{\mathbf{k},\mathbf{q}}+\Omega_0)\Big\{g_{\mathbf{k}}^{\mathbf{q}}+\tfrac{1}{2}\sum_{\mathbf{k'}}f_e(\mathbf{k},\mathbf{k'})\,g_{\mathbf{k'}}^{\mathbf{q}}\,\delta(\epsilon_{\mathbf{k'}}-E_{\mathrm{F}})\Big\}$$
$$= \text{damping and excitation terms.} \quad (10.45)$$

The spin magnetization $M(\mathbf{q},\omega)$ is obtained by summing $g_{\mathbf{k}}^{\mathbf{q}}$ over all the values of $\mathbf{k}$:

$$M(\mathbf{q},\omega) = \sum_{\mathbf{k}} g_{\mathbf{k}}^{\mathbf{q}}\,\delta(\epsilon_{\mathbf{k}}-E_{\mathrm{F}}). \quad (10.46)$$

Thus the amplitude of the magnetization is proportional to the amplitude of this collective excitation, obtained when all the $g_{\mathbf{k}}^{\mathbf{q}}$ have comparable magnitudes.

Using (10.45) and (10.46), the variation of $M(\mathbf{q},\omega)$ is obtained. In the limit of very small $\mathbf{q}$ the equation may be written

$$iM(\mathbf{q},\omega)(\Omega_0-\omega)-\sum_{\mathbf{k},\mathbf{k'}}\{\tfrac{1}{2}f_e(\mathbf{k},\mathbf{k'})\,g_{\mathbf{k'}}^{0}\,\delta(\epsilon_{\mathbf{k'}}-E_{\mathrm{F}})\}$$
$$= \text{damping+excitation.} \quad (10.47)$$

If the summation over $\mathbf{k}$ is performed using (3.23), we find that the eigenvalue is $\omega = \Omega_0(1+B_0) = \omega_0$. This important result was expected. For low values of $\mathbf{q}$, $M(\mathbf{q},\omega)$ tends towards the uniform spin magnetization and this quantity is not affected by an exchange interaction. (For a uniform excitation the spins of all the quasi-particles are turned by the same angle and the exchange energy does not change.)

The complete spectrum of the excitations is shown in Fig. 10.4. In all these considerations the Lorentz force was neglected.

For non-zero values of the wave vector $\mathbf{q}$, the energy of the collective excitation, called the spin wave mode, varies as q^2. It can be shown that the susceptibility is a simple function of the energy of the spin wave mode, namely

$$\chi_S(\mathbf{q},\omega) = \chi_S\,\frac{\omega_0}{\omega(q)-\omega}. \quad (10.48)$$

χ_S is the static spin susceptibility and $\hbar\,\omega(\mathbf{q})$ the energy of the spin wave mode. Wolff and Platzmann[179] have calculated $\omega(\mathbf{q})$ up to the second order in $\mathbf{q}$ and obtain the result

$$\omega(q) = \omega_0-\tfrac{1}{3}q^2\,v_{\mathrm{F}}^2(1+B_0)(1+B_1)\Big(\omega_0+\frac{i}{\tau}-\omega_0\frac{1+B_1}{1+B_0}\Big)\times$$
$$\times\Bigg\{\frac{\sin^2\Delta}{\omega_c^2(1+B_1)^2-\{\omega_0(B_0-B_1)/(1+B_0)+i/\tau\}^2}-$$
$$-\frac{\cos^2\Delta}{\{\omega_0(B_0-B_1)/(1+B_0)+i/\tau\}^2}\Bigg\}. \quad (10.49)$$

ω_c is the cyclotron resonance frequency and B_1 is another function of f_e defined as follows:

$$B_1 = \tfrac{1}{6} \sum_{\mathbf{k'}} \cos\theta_{\mathbf{k,k'}} f_e(\mathbf{k,k'})\,\delta(\epsilon_{\mathbf{k'}} - E_F).$$

Δ is the angle between the magnetic field and the z-axis (here the Lorentz force has been taken into account). The frequency $\omega(q)$ is a complex quantity, i.e. both a dispersive and an absorptive component are present. The imaginary part arises from the finite lifetime τ.

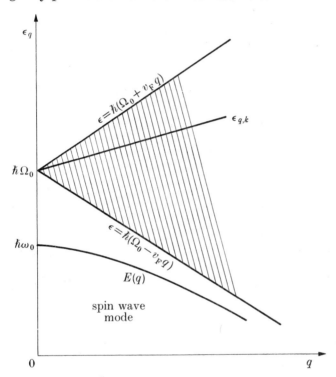

FIG. 10.4. The spectrum of the transverse elementary spin excitations in a metal.

The two equations (10.48) and (10.49) include many of the results discussed at the beginning of this chapter. The behaviour changes drastically according to the relative magnitude of the real and imaginary parts of $\omega(q)$. For instance, if the interaction between electrons is weak, or more precisely if $\omega_0 \tau \ll B_0, B_1$, i.e. the inverse lifetime is large compared with the exchange frequency, eqn (10.49) becomes

$$\omega(q) = \omega_0 - \frac{i}{3} q^2 v_F^2 \frac{1}{\tau}\left(\frac{\sin^2\Delta}{\omega_c^2 + (1/\tau)^2} + \frac{\cos^2\Delta}{(1/\tau)^2}\right). \tag{10.50}$$

N

The second term on the right-hand side is pure imaginary and the equation may be interpreted by assuming that the magnetization obeys a diffusion equation whose diffusion constant varies with the direction of the field as predicted by eqn (10.16). All the results derived in section 2 of this chapter are a consequence of this equation.

When the damping is small, however, new results are obtained. Let us assume now that $1/\tau$ is very small in that $\omega_0 \tau (B_0 - B_1) \gg 1$. The spin wave energy then has the value

$$\omega(q) = \omega_0 - \tfrac{1}{3} q^2 v_{\mathrm{F}}^2 \frac{\omega_0 (B_0 - B_1)}{1 + B_0} \times$$

$$\times \left\{ \frac{\sin^2 \Delta}{\omega_c^2 (1 + B_1)^2 - \omega_0^2 \{(B_0 - B_1)^2 / (1 + B_0)^2\}} - \frac{\cos^2 \Delta}{\omega_0^2 \{(B_0 - B_1)^2 / (1 + B_0)^2\}} \right\},$$

$$(10.51)$$

where $\omega(q)$ is a real quantity. $\chi(q, \omega)$ is also real and, using (10.44), the shape of the transmitted signal may be calculated. We find that the signal is large when ω is equal to the energies of the following spin wave modes:

$$\omega = \omega \left(q = \frac{n\pi}{l} \right),$$

where n is an integer. Several peaks are observed corresponding to the energies of these excitations. These excitations are modes having a large amplitude for the magnetization at the two ends of the sample, but the magnetization varies and presents n nodes along the z-axis (see Fig. 10.5). The position of the satellite peaks is a function of B_0, B_1, Δ, and $1/\tau$. Fig. 10.6 shows an experimental spectrum which is compared with the theoretical prediction. The agreement is rather good.

The position of the peaks is a function of Δ (see Fig. 10.7), for a certain value of which, Δ_c, the peaks coincide with the main line ($\omega = \omega_0$). From the experimental value of Δ_c, the ratio B_0/B_1 has been measured with a high degree of accuracy.

The following results are obtained:

for sodium $B_0 = -0{\cdot}18 \pm 0{\cdot}03$, $B_1 = 0{\cdot}05 \pm 0{\cdot}04$;

for potassium $B_0 = -0{\cdot}28 \pm 0{\cdot}10$, $B_1 = -0{\cdot}06 \pm 0{\cdot}15$.

The value of B_0 for sodium is in good agreement with the value deduced from the measurement of χ_S by Slichter[41] but in substantial disagreement with the Schumacher result;[40] see Table 10.

6. Electronic resonances in transitional alloys

6.1. *Introduction*

In transition metals or alloys two kinds of electronic resonances may be observed. It is sometimes possible to observe the resonance due to conduction electrons as, for example, in alloys of copper with a transition element (Mn, Cr, or Fe) or in silver alloys.

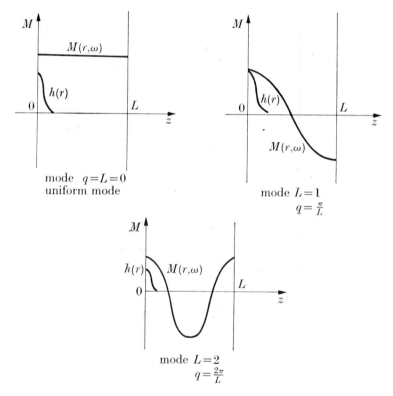

FIG. 10.5. Variation of the magnetization across the plate for the first spin-wave modes.

In these alloys the fact that the resonance is due to conduction electrons is determined by the analysis of the line shape observed using the transmission technique.

In other cases, where localized moments exist, it is possible to observe the electronic resonance of these centres. We shall discuss first the properties of localized impurities.

6.2. *Resonance of localized electronic spins in metals*

The properties of the electronic resonance of impurities present some similarity with the nuclear resonance properties in a metal (for a review of this subject, see Ref. 199).

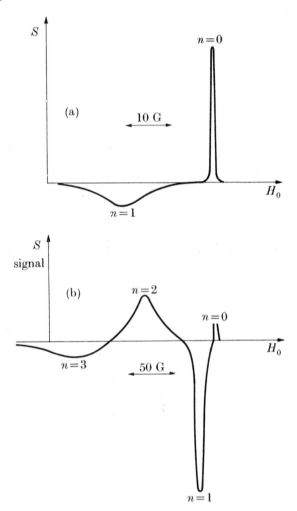

Fig. 10.6. Typical spin-wave signals in sodium; in (b) the gain and field sweep are expanded.[194,195]

There are, however, some important differences which will be enumerated now. The perturbation produced by the impurity is large. The form of the interaction between the conduction electrons and the impurity localized spin is not easy to establish. Also the resonance frequencies of the two systems are very close and this fact complicates the interpretation of the observed shifts.

The simplest impurities in metals are Mn^{++}, Gd^{+++}, and Eu^{++}; these three ions are characterized by having a half-filled internal shell ($3d$ or $4f$). In an ionic crystal these ions have a ground state without orbital

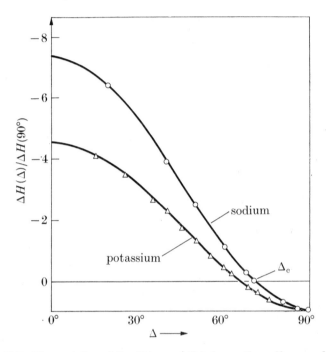

FIG. 10.7. The variation of the distance ΔH between the uniform mode and the first spin wave mode as a function of the angle Δ between the magnetic field and the normal to the plate.[194]

moment (^6S or ^8S state). The experiments show that in a metal the magnetic properties of these ions are not too affected by the conduction electrons. For example, the g factors are only slightly changed.

As discussed in Chapter 8, these spins are coupled to the conduction electrons by an exchange interaction that, for simplicity, will be assumed to have the form given by eqn (8.9):

$$\mathscr{H} = J_{sd}\,\mathbf{S}_d \cdot \mathbf{S}_F(R). \tag{8.9}$$

This interaction is similar to the hyperfine contact coupling and will have similar effects on the spin of the ion. A shift of the electronic resonance line will be observed.

However, as the two resonance frequencies are nearly equal when we observe the resonance of the localized spins, the resonance of the conduction electrons is also excited and we have to find the eigenfrequencies of this system of coupled spins.[199, 200] The resonance frequency is not only a function of the two frequencies but also of J_{sd} and of the damping of the two systems. When the damping of the conduction electronic resonance is very short we find a shift for the localized spins given by the average value of the field due to the conduction electrons. The complete set of coupled equations will be given in the next section. The average electronic field using (8.9) is proportional to $\chi_S J_{sd}$ and, as in the Knight-shift discussion, the exchange constant may be estimated if χ_S is known.

The coupling will also produce a relaxation of the localized spins, the rate for which may be calculated exactly as the nuclear relaxation rate and will vary as $k_B T J_{sd}^2$. This rate is usually very large because J_{sd} is large and a broadening of the resonance line varying as $k_B T$ is observed in the high-temperature range. But here too the same interaction will contribute to the relaxation rate of the conduction electrons. (This term will give a contribution to the spin-flip scattering cross-section.)

In some experiments the resonance of Mn^{++} (or of another localized spin ion) is observed while adding to the metal another magnetic impurity. Here the situation presents a similarity with the study of nuclear resonance in a magnetic alloy. The change of the electronic magnetization around the impurity will broaden the electronic resonance line of Mn^{++}. This method is convenient for measuring the value of the exchange interactions of impurities that do not give an observable electronic resonance signal (the shifts produced by these ions are called indirect g shifts).

6.3. *Resonances using the transmission technique*

Recent experiments were performed using the transmission technique in copper or silver alloys with transition elements (Cr, Fe, or Mn). On adding these impurities the width, shape, and position of the transmitted signal in copper are changed.[65, 66]

The theoretical model used to explain the results is rather simple. One assumes that two types of magnetization are present, $\mathbf{M_s}$ and $\mathbf{M_d}$, coupled by a local exchange interaction. The quantity $\mathbf{M_s}$, which is the magnetization of the conduction electrons, obeys a diffusion equation (for $\mathbf{M_d}$ which is localized $D = 0$). Finally, as explained in the previous section, there are cross-relaxation terms that tend to equalize the temperatures of the two systems. The equations are as follows :[201]

$$
\left.
\begin{aligned}
\frac{\mathrm{d}\mathbf{M_s}}{\mathrm{d}t} &= \gamma_s\,\mathbf{M_s}\wedge\mathbf{H_0}+\lambda\gamma_s\,\mathbf{M_s}\wedge\mathbf{M_d}-\frac{1}{T_{sd}}\mathbf{M_s}+ \\
&\quad +\frac{1}{T_{ds}}\mathbf{M_d}-\frac{1}{T_{sl}}(\mathbf{M_s}-\mathbf{M_s^0})+D\nabla^2\mathbf{M_s} \\
\frac{\mathrm{d}\mathbf{M_d}}{\mathrm{d}t} &= \gamma_d\,\mathbf{M_d}\wedge\mathbf{H_0}+\lambda\gamma_d\,\mathbf{M_d}\wedge\mathbf{M_s}+\frac{1}{T_{sd}}\mathbf{M_s}- \\
&\quad -\frac{1}{T_{ds}}\mathbf{M_d}-\frac{1}{T_{dl}}(\mathbf{M_d}-\mathbf{M_d^0})
\end{aligned}
\right\}. \quad (10.52)
$$

The parameter λ is proportional to the value of the exchange coupling J_{sd}. The cross-relaxation rates, T_{sd} and T_{ds}, are not independent because in an equilibrium situation the magnetization of the two systems are constant and known. Using (10.52) we get the relation

$$
\frac{M_s^0}{T_{sd}}=\frac{M_d^0}{T_{ds}} \quad\text{or}\quad \frac{T_{ds}}{T_{sd}}=\frac{\chi_d}{\chi_s}=\chi_r, \quad (10.53)
$$

T_{ds} is the relaxation rate for the magnetization of Mn^{++} as defined in the previous section, T_{sd} is the broadening of the line of the conduction electron due to the exchange interaction with the impurities. (T_{ds} varies as $J_{sd}^2\,k_B\,T$, whereas T_{sd} does not vary with the temperature.) T_{sl} and T_{dl} are relaxation times due to the direct coupling between the systems and the lattice and χ_d and χ_s are the susceptibilities.

By adding the Maxwell equations to eqn (10.52), the transmitted signal may be calculated. But this system may also be used to describe situations where the resonance of conduction electrons is not observed (due for instance to a too short value for T_{sl}) but where the resonance of the localized system is observed as discussed in the previous section.

An important parameter for discussing the results is the ratio χ_r of the susceptibilities. Whenever χ_r is smaller than 1, the properties of the conduction electrons will dominate and the signal will behave almost as in pure copper; when χ_r is large the resonance will be dominated by the properties of the localized centres. χ_r may be varied by changing the impurity concentration or more simply by varying the temperature.

When the exchange is large, more precisely when the condition

$$
|\omega_s-\omega_d| < \left|\lambda\chi_s\,\omega_s+\frac{i}{T_{ds}}\right|
$$

is fulfilled, the calculation of the transmitted signal becomes much simpler. The results are as follows. The signal is the same as for a system

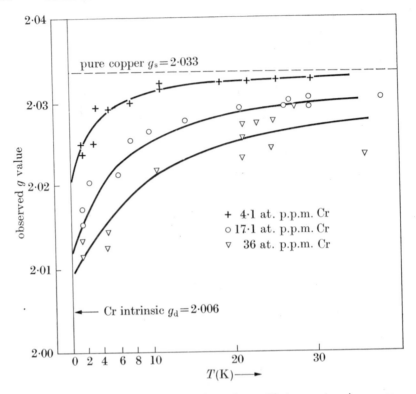

FIG. 10.8. Variation of the observed g values with temperature in copper–chromium alloys. The solid lines represent the theoretical predictions.[66,190]

of conduction electrons providing a suitable redefinition of the parameters $\omega_{\rm s}$, $1/T_1$, and D. These quantities are replaced by effective parameters $\omega_{\rm eff}$, $(1/T_1)_{\rm eff}$, and $D_{\rm eff}$. They have the values

$$\left.\begin{aligned}
\omega_{\rm eff} &= \omega_{\rm s} + \frac{(\omega_{\rm d} - \omega_{\rm s})\chi_{\rm r}}{1 + \chi_{\rm r}} \\[2mm]
D_{\rm eff} &= D\frac{1}{1 + \chi_{\rm r}} \\[2mm]
\left(\frac{1}{T_1}\right)_{\rm eff} &= \frac{1}{T_{\rm sl}}\frac{1}{1 + \chi_{\rm r}} + \frac{1}{T_{\rm dl}}\frac{\chi_{\rm r}}{1 + \chi_{\rm r}}
\end{aligned}\right\}. \qquad (10.54)$$

These three results have a simple physical interpretation. $\omega_{\rm eff}$ is the average resonance frequency of the two systems with a weight proportional to their equilibrium magnetization. (This result may be obtained by solving (10.52) keeping in the equations only the first two terms of the right-hand side.)

The effective line width and the effective diffusion constant are also averages of the line width and of the diffusion constants ($D_s = D$, $D_d = 0$).

Detailed experiments were carried out in chromium- and manganese-doped copper.

The observed line width decreases if the temperature increases up to 25 K (for higher temperature a broadening due to the phonons is observed). This effect arises because $1/T_{dl}$ is very large and the second term in $(1/T_1)_{eff}$ decreases if the temperature increases. From the measurements of ω_{eff} as a function of the temperature, χ_r may be deduced (see Fig. 10.8); knowing χ_r, T_{dl} and T_{sl} are estimated.

From knowledge of χ_r the value of the susceptibility for pure copper is deduced. For a metal having a spherical Fermi surface (a questionable approximation for copper) eqn (3.24) may be used and B_0 is known. In contradistinction to the results found for sodium or potassium, B_0 is found to be positive. From the variation of T_{sl} with the concentration of impurities the cross-section for spin flip scattering may also be estimated.

There is no explanation for the mechanism giving rise to the relaxation time T_{dl}.

APPENDIX 1

RELATIONS BETWEEN SUSCEPTIBILITIES AND CORRELATION FUNCTIONS

1. General relations

THESE relations are discussed in many references and various approaches are used, see References 9, 10, 22, 23, and 202.

Let us first define the time-variable susceptibility in a rather general way. We start with a system described by a Hamiltonian $\mathscr{H}$ and consider two operators A and B. For simplicity it will be assumed that the thermal average values $\langle A \rangle$ and $\langle B \rangle$ are equal to zero and as usual we write

$$\langle A \rangle = \mathrm{Tr}(\rho A) \quad \text{with} \quad \rho = \mathrm{e}^{-\mathscr{H}/k_B T} / \mathrm{Tr}(\mathrm{e}^{-\mathscr{H}/k_B T}).$$

The system is perturbed by the Hamiltonian

$$\mathscr{H}'(t) = -A F(t),$$

where $F(t)$ is a c-number function of the time variable. The problem is to calculate the new average value of the operator B; this quantity will be called $\Delta B(t)$:

$$\Delta B(t) = \mathrm{Tr}\{\rho + \Delta\rho(t),\, B\},$$

where $\Delta\rho(t)$ is the change in the density matrix due to the perturbation. The perturbation is assumed to be weak and we look only for the terms linear in $F(t)$. To calculate $\Delta\rho(t)$ and $\Delta B(t)$ we start from the master equation for the density matrix:

$$\frac{\mathrm{d}}{\mathrm{d}t}\{\rho + \Delta\rho(t)\} = -\frac{\mathrm{i}}{\hbar}\{\mathscr{H} + \mathscr{H}'(t),\, \rho + \Delta\rho(t)\}$$

or

$$\frac{\mathrm{d}}{\mathrm{d}t}\{\Delta\rho(t)\} = -\frac{\mathrm{i}}{\hbar}\{\mathscr{H},\, \Delta\rho(t)\} - \frac{\mathrm{i}}{\hbar}[\mathscr{H}',\, \rho],$$

keeping only the first-order terms. This equation is easily integrated with the help of the intermediate variable $\Delta\tilde{\rho} = \mathrm{e}^{-\mathrm{i}\mathscr{H}t/\hbar}(\Delta\rho)\mathrm{e}^{\mathrm{i}\mathscr{H}t/\hbar}$. The result is

$$\Delta\rho(t) = \frac{\mathrm{i}}{\hbar} \int_{-\infty}^{t} \mathrm{e}^{-\mathrm{i}(t-t')\mathscr{H}/\hbar}[A, \rho]\mathrm{e}^{\mathrm{i}(t-t')\mathscr{H}/\hbar} F(t')\, \mathrm{d}t';$$

thus

$$\Delta B(t) = \frac{\mathrm{i}}{\hbar}\mathrm{Tr} \int_{-\infty}^{t} [A, \rho]B(t-t')F(t')\, \mathrm{d}t'. \tag{A1.1}$$

$B(t)$ is the operator B in the Heisenberg representation

$$B(t) = \mathrm{e}^{\mathrm{i}\mathscr{H}t/\hbar} B\, \mathrm{e}^{-\mathrm{i}\mathscr{H}t/\hbar}.$$

Let us now consider the time variation $F = F_0\,\mathrm{e}^{\mathrm{i}\omega t}$. The susceptibility $\chi_{BA}(\omega)$ is defined by the relation

$$\Delta B(t) = F_0\,\mathrm{e}^{\mathrm{i}\omega t}\chi_{BA}(\omega).$$

In the linear approximation, if the excitation is periodic the response must also be periodic.

Using (A1.1) we obtain for $\chi_{BA}(\omega)$

$$\chi_{BA}(\omega) = \frac{i}{\hbar} \text{Tr} \int_{-\infty}^{t} [A, \rho] B(t-t') e^{i\omega(t'-t)} \, dt',$$

which can be written as

$$\chi_{BA}(\omega) = \frac{i}{\hbar} \int_{0}^{\infty} \langle [B(\tau), A] \rangle e^{-i\omega\tau} \, d\tau. \tag{A1.2}$$

This result assumes that the integral is not divergent but it may happen that the quantity $(1/\hbar)\langle [B(\tau), A] \rangle$ does not tend towards zero for large values of τ and the result will become meaningless. The susceptibility may be more safely defined by the relation

$$\chi_{BA}(\omega) = \lim_{\epsilon \to 0} \frac{i}{\hbar} \int_{0}^{\infty} \langle [B(\tau), A] \rangle e^{-i\omega\tau - \epsilon\tau} \, d\tau. \tag{A1.3}$$

(For a more complete discussion of this point, see Ref. 202.)

Using this definition there is no ambiguity about the sign of ϵ; for example, if we calculate $\chi_{BA}(-\omega)$ the sign of ϵ does not change.

This equation looks rather similar to the Fourier transform of a correlation function but differs in two respects; first, the integration is restricted to positive values of the time and secondly the quantity to integrate is not a simple product but a commutator.

Let us consider another similar susceptibility $\chi_{AB}(-\omega)$. Using (A1.3), we have

$$\chi_{AB}(-\omega) = \lim_{\epsilon \to 0} \frac{i}{\hbar} \int_{0}^{\infty} \langle [A(\tau), B] \rangle e^{i\omega\tau - \epsilon\tau} \, d\tau$$

$$= -\lim_{\epsilon \to 0} \frac{i}{\hbar} \int_{-\infty}^{0} \langle [B, A(-\tau)] \rangle e^{-i\omega\tau + \epsilon\tau} \, d\tau.$$

As it can be shown that $\langle A(-\tau)B \rangle = \langle AB(\tau) \rangle$, we get

$$\chi_{BA}(\omega) - \chi_{AB}(-\omega) = \lim_{\epsilon \to 0} \frac{i}{\hbar} \int_{-\infty}^{+\infty} \langle [B(\tau), A] \rangle e^{-i\omega\tau - \epsilon|\tau|} \, d\tau. \tag{A1.4}$$

We look for a relation between this combination of susceptibilities and the Fourier transform of the correlation function defined as

$$J_{BA}(\omega) = \frac{1}{2\pi} \int_{-\infty}^{+\infty} \langle B(\tau)A \rangle e^{-i\omega\tau - \epsilon|\tau|} \, d\tau.$$

Let us calculate the trace. The energy levels will be called E_n and E_m, and A_{nm} and B_{mn} are the matrix elements of the operators A and B (we shall neglect the term $e^{-\epsilon|\tau|}$):

$$J_{BA}(\omega) = \frac{1}{2\pi} \sum_{n,m} \int_{-\infty}^{+\infty} d\tau \, e^{-i\omega\tau} e^{i(E_n - E_m)\tau/\hbar} e^{-E_n/k_B T} A_{mn} B_{nm},$$

and for the susceptibility

$$\chi_{BA}(\omega) - \chi_{AB}(-\omega) = \frac{2\pi i}{\hbar} J_{BA}(\omega) - \sum_{n,m} \frac{i}{\hbar} \int_{-\infty}^{+\infty} e^{-i\omega\tau} \, d\tau \, e^{-E_m/k_B T} e^{i(E_n - E_m)\tau/\hbar} A_{nm} B_{mn}.$$

As the integration over τ gives the delta function $\delta(E_n - E_m - \hbar\omega)$, the two integrals differ only by the constant factor

$$e^{(E_n - E_m)/k_B T} = e^{\hbar\omega/k_B T}.$$

Thus
$$J_{BA}(\omega) = \frac{\hbar}{2\pi i} \frac{\chi_{BA}(\omega) - \chi_{AB}(-\omega)}{1 - e^{\hbar\omega/k_B T}}. \tag{A1.5}$$

When the operators A and B are Hermitian, because to a real excitation we must have a real response, the relation $\chi_{BA}(\omega) = \chi_{BA}^*(-\omega)$ is obtained. In the special case $A = B$ (and A Hermitian) $J_{BA}(\omega)$ is proportional to the imaginary part of the susceptibility.

2. Application to spin susceptibilities

If we go back to the relaxation calculation of Chapter 4 (eqn (4.40)), the considerations about $1/T_1$ are obtained using (A1.5) with $A = S_-(R_i)$ and $B = S_+(R_i)$. Again, if the Zeeman energy is neglected, all these local susceptibilities are functions of χ_{BA} with $A = B = S_z(R_i)$,

$$\chi_S(R_i, R_i, \omega) = \chi_{BA}(\omega), \qquad A = B = S_z(R_i).$$

In Chapter 3, we considered the susceptibility $\chi_S(\mathbf{q}, \omega)$. This quantity is obtained using $A = S_{\mathbf{q}z}$, $B = S_{-\mathbf{q}z}$. By requiring that a real perturbation produces a real response, we obtained the relation

$$\chi_S(\mathbf{q}, \omega) = \chi_S^*(-\mathbf{q}, -\omega).$$

APPENDIX 2

NUCLEAR RELAXATION IN TRANSITION METALS

1. Orbital and dipolar relaxations[19]

IN Chapter 4, the principle of the calculation of the orbital and dipolar relaxation rates was discussed. These considerations will be applied now to the case of d-band electrons in a transition metal.

The d-electron wave functions are usually calculated using the tight binding approximation. Let us first describe this method.

1.1. Tight binding wave functions

The wave functions are built using linear combinations of localized atomic wave functions:

$$\Phi_{\mathbf{k},n}(\mathbf{r}) = \sum_{m,\mathbf{R}_i} c_{\mathbf{k},n}^m a_m(\mathbf{r}-\mathbf{R}_i) e^{i\mathbf{k}.\mathbf{R}_i}, \qquad (A2.1)$$

where $a_m(\mathbf{r}-\mathbf{R}_i)$ is an atomic wave function for the atom centred at the point $\mathbf{R}_i$, n is the band index, and m an index that labels all the atomic wave functions. If the overlap between atomic wave functions belonging to different atoms is neglected, the two sets of wave functions, $\{\Phi_{\mathbf{k},n}\}$ and $\{a_m(\mathbf{r}-\mathbf{R}_i)\}$, are complete and orthogonal and the coefficients $c_{\mathbf{k},n}^m$ define a unitary transformation.

For d-band electrons in a transition metal the approximation will be to consider in the expansion (A2.1), only the d-wave functions $a_m^d(\mathbf{r}-\mathbf{R}_i)$ (and eventually an s-part $a^s(\mathbf{r}-\mathbf{R}_i)$ if the s–d mixing is taken into account). The index m is now labelling the orbital degeneracy of the d-wave functions.

It is convenient to choose these functions in such a way that they form the basis of an irreducible representation of the point group of the system. For cubic symmetry the d-wave functions span the bases of two representations Γ_3 (with a double degeneracy) and Γ_5 (with a triple degeneracy). For a simple metal the same method may be used in the expansion (A2.1), but only the s- and p-wave functions are considered.

1.2. Orbital and dipolar rates

These rates are calculated using eqn (4.29) and a similar equation for the dipolar rate. The advantage of using the tight binding wave functions is that the calculation of the matrix elements of the orbital (or dipolar) hyperfine field is done as for an atom (or an ion) in an insulator. The evaluation of these quantities is therefore similar to the calculation of the hyperfine effects in paramagnetic resonance.[4] Here also it is quite important to take into account the invariance properties of the system to the operations of the appropriate point group. Using this property we find a relation between the coefficients $c_{\mathbf{k},n}^m$:[19]

$$\sum_{\mathbf{k},n} c_{\mathbf{k},n}^{*m} c_{\mathbf{k},n}^{m'} \delta(\epsilon_{\mathbf{k},n}-E_{\mathrm{F}}) = \delta_{mm'} C_m g(E_{\mathrm{F}}), \qquad (A2.2)$$

where C_m is a constant that is a function of the dimension of the representation

defined by the atomic functions; for example, for d-wave functions we have two
constants, C_{Γ_3} and C_{Γ_5}, with the relation

$$2C_{\Gamma_3} + 3C_{\Gamma_5} = 5.$$

Detailed calculations of the relaxation rates for cubic symmetry are found in
Reference 19 and for hexagonal symmetry in Reference 203.

For cubic symmetry and pure d-functions, if p is the fractional admixture of
Γ_5 wave functions for wave functions at the Fermi level the following orbital
relaxation rate is found:

$$\left(\frac{1}{T_1}\right)_\sigma = \frac{4\pi}{\hbar}(\gamma_n\hbar g\beta)^2 k_B T g_d(E_F)\left\langle\frac{1}{r^3}\right\rangle^2 6p(2-\tfrac{5}{3}p). \tag{A2.3}$$

The dipolar rate is found to be much smaller (except when p is very small). $\langle 1/r^3\rangle$
is the average value of $1/r^3$ using the radial part of the d-wave functions.

The orbital rate vanishes when $p = 0$ for, as is well known, there is no matrix
element for the orbital momentum operator between wave functions of the mani-
fold Γ_3.

1.3. *Anisotropy of orbital and dipolar relaxations*

As already stated in Chapter 4, these rates may depend upon the orientation
of the magnetic field with respect to the crystalline axis.

For calculating this anisotropy it is more convenient to take a frame of reference
related to the crystalline axis of symmetry and not, as in eqn (9.29), to the direction
of the external field.

If the hyperfine interaction is written as

$$\mathscr{H}_0 = -\hbar\gamma_n\mathbf{I}.\mathbf{H}_{\mathrm{orb}} \quad (\text{or } -\hbar\gamma_n\mathbf{I}.\mathbf{H}_{\mathrm{dip}})$$

we have to calculate the matrix elements of the operator $H_{\mathrm{orb}\,x}+iH_{\mathrm{orb}\,y}$ using the
former axes. If these quantities are expressed in terms of $H_{\mathrm{orb}\,X}$, $H_{\mathrm{orb}\,Y}$, and $H_{\mathrm{orb}\,Z}$
in the new system, they are linear functions of the direction cosines (α,β,γ) of the
former z-axis with respect to the new axis. Therefore the rate that involves the
square of the matrix elements must be a quadratic function of α, β, γ. This result
is also valid for the dipolar rate provided the Zeeman energies are neglected.

Consequently for cubic symmetry the rates are isotropic (if the symmetry is
cubic the possible anisotropy involves a function of α, β, and γ of order 4). This
result was obtained by Obata[19] in the special case of tight binding p- or d-wave
functions. (In the calculation of the dipolar rate Obata found an anisotropy for
some of the matrix elements: this result happens because he does not rotate the
spin part of the dipolar field; however, as expected, the total rate is isotropic.)

For symmetries lower than cubic an anisotropy may appear and indeed was
observed in the hexagonal transition metal scandium.[204]

2. Core–polarization relaxation[109]

2.1. *Introduction*

The physical idea of the core-polarization effect was introduced in Chapter 2.
However, the model we described is unable to predict the core-polarization relaxa-
tion and this effect has to be more deeply discussed. The method introduced in
Chapter 2, usually called the 'unrestricted Hartree–Fock method' (or more briefly
U.H.F.) consists of assuming that the radial parts of the s-wave functions having

different spin orientations are different. Let us first discuss a simple model of a three-electron system $(1s)^2(d)$ (the orbital degeneracy of the d-wave function will be neglected). The zero-order wave functions are the following Slater determinants:

$$\Psi_\uparrow^0 = (1s\uparrow)(1s\downarrow)(d\uparrow)$$

and

$$\Psi_\downarrow^0 = (1s\uparrow)(1s\downarrow)(d\downarrow).$$

In the U.H.F. method the perturbed wave functions are

$$\Psi_\uparrow = (1s\uparrow)(1s+\alpha ps\downarrow)(d\uparrow)$$

and

$$\Psi_\downarrow = (1s+\alpha ps\uparrow)(1s\downarrow)(d\downarrow), \tag{A2.4}$$

where ps is another s-orbital and α is the admixture coefficient. With these wave functions the operator $S_z(\mathbf{R}_i)$ (where $\mathbf{R}_i$ is the nucleus position) has an average value $\langle \psi_\uparrow | S_z(\mathbf{R}_i) | \psi_\uparrow \rangle$, which is proportional to the product of the s-wave functions $\alpha a_{1s}(0) a_{ps}(0)$. However, the wave functions (A2.4) present a serious disadvantage; they are not eigenfunctions of the total spin operator $\mathbf{S}$. This fact is proved by the following result. If we calculate a non-diagonal element of the spin operator $\langle \psi_\downarrow | S_+(\mathbf{R}_i) | \psi_\uparrow \rangle$ we find that this quantity is equal to zero, although rotation invariance arguments predict a value equal to $2\langle \psi_\uparrow | S_z(\mathbf{R}_i) | \psi_\uparrow \rangle$. In the other method, called the 'configuration interaction method', the wave function of the ground state Ψ^0 is admixed with the excited configurations $(1s, ps, d)$ having the same quantum numbers (thus $\mathbf{S}$ will still be a good quantum number). Such a function will differ from the U.H.F. functions (A2.4) only by a correction term; we find[4]

$$\Psi_\uparrow = (1s\uparrow)(1s+\alpha ps\downarrow)(d\uparrow) - (1s\uparrow)(\alpha ps\uparrow)(d\downarrow).$$

The presence of the last term does not change the average value of $S_z(\mathbf{R}_i)$ but restores the expected value for $\langle \Psi_\downarrow | S_+(\mathbf{R}_i) | \Psi_\uparrow \rangle$.

2.2. Core polarization in a metal with a d-band

The preceding considerations made for an atom are now extended to the case of a metal. For simplicity it will be assumed that only one internal s-shell is present.

The zero-order wave function for a situation where a d-electron of wave vector $\mathbf{k}$ is above the Fermi energy will be written as

$$\psi_{k\uparrow}^0 = [\ \ ]...[1s, \mathbf{K}, \uparrow][1s, \mathbf{K}, \downarrow]...[\mathbf{k}, \uparrow].$$

The state $[1s, \mathbf{K}]$ is an admixture of the individual $1s$-orbital, defined as follows:

$$a_{1s}(\mathbf{r} - \mathbf{R}_i) = \sum_{\mathbf{K}} e^{-i\mathbf{K}\cdot\mathbf{R}_i} \phi_{s\mathbf{K}}(r); \tag{A2.5}$$

the summation over $\mathbf{K}$ is restricted to the first Brillouin zone.

As in the atomic case, this zero-order wave function will be admixed with wave functions of the form

$$\Psi_{k\uparrow}^1 = [...][1s, \mathbf{K}, \uparrow][ps, \mathbf{K}, \downarrow]...[\mathbf{k}, \uparrow]$$

and

$$\Psi_{k\uparrow,q}^3 = [...][1s, \mathbf{K}, \uparrow][ps, \mathbf{K}-\mathbf{q}, \uparrow]...[\mathbf{k}+\mathbf{q}, \downarrow]. \tag{A2.6}$$

The admixture with $\Psi_{k\uparrow}$ will give the core-polarization contribution to the Knight shift, whereas the core-polarization relaxation rate is due to the wave function $\Psi_{k\uparrow,q}^3$.

We are now able to calculate the two effects. The calculation proceeds by the following steps: First, the coefficients of admixture between the wave functions are calculated. Let us call $e_1(p, \mathbf{K}, 0, \mathbf{k})$ the coefficient for the admixture between $\Psi^1_{\mathbf{k}\uparrow}$ and $\Psi^0_{\mathbf{k}\uparrow}$; this quantity has the value

$$e_1(p, \mathbf{K}, 0, \mathbf{k}) = \frac{(ps\mathbf{K}, \mathbf{k}|g|\mathbf{k}, 1s\mathbf{K})}{E_{1s} - E_{ps}},$$

where g is the coulomb interaction. A similar definition is obtained for the admixture coefficient $e_3(p, \mathbf{K}, \mathbf{q}, \mathbf{k})$ between $\Psi^0_{\mathbf{k}\uparrow}$ and $\Psi^3_{\mathbf{k}\uparrow,\mathbf{q}}$. Then the matrix elements of $S_z(\mathbf{R}_i)$ and $S_+(\mathbf{R}_i)$ are obtained:

$$\langle\Psi_{\mathbf{k}\uparrow}|S_z(\mathbf{R}_i)|\Psi_{\mathbf{k}\uparrow}\rangle$$

$$= \tfrac{1}{2}\Big\{|\phi^d_{\mathbf{k}}(\mathbf{R}_i)|^2 + \sum_{p,\mathbf{K}} \{e_1(p, \mathbf{K}, 0, \mathbf{k})\phi^*_{1s\mathbf{K}}(\mathbf{R}_i)\phi_{ps\mathbf{K}}(\mathbf{R}_i)\} + \text{complex conjugate}\Big\} \quad (A2.7)$$

and

$$\langle\Psi_{\mathbf{k}+\mathbf{q}\downarrow}|S_+(\mathbf{R}_i)|\Psi_{\mathbf{k}\uparrow}\rangle = \phi^{*d}_{\mathbf{k}+\mathbf{q}}(\mathbf{R}_i)\phi^d_{\mathbf{k}}(\mathbf{R}_i) + \sum_{p,\mathbf{K}} e_3(p, \mathbf{K}, \mathbf{q}, \mathbf{k})\phi^*_{1s\mathbf{K}}(\mathbf{R}_i)\phi_{ps\mathbf{K}-\mathbf{q}}(\mathbf{R}_i) +$$

$$+ e_3(p, \mathbf{K}, -\mathbf{q}, \mathbf{k}+\mathbf{q})\phi_{1s\mathbf{K}}(\mathbf{R}_i)\phi^*_{ps.\mathbf{K}+\mathbf{q}}(\mathbf{R}_i). \quad (A2.7')$$

Now the functions $\phi_{1s\mathbf{K}}$ and $\phi_{ps\mathbf{K}}$ are expressed in terms of the localized wave functions a_{1s}, using (A2.5). Similarly, $\phi^d_{\mathbf{k}}$ is calculated using the tight binding expansion (A2.1) keeping the d- and s-parts and using the basis functions a^d_m of irreducible representations of the relevant point group. In the product of the s-wave functions we consider only the a_s functions belonging to the same atom. This function is also used for calculating the coefficients e_1 and e_3. The matrix elements are as follows:

$$\langle\Psi_{\mathbf{k}\uparrow}|S_z(\mathbf{R}_i)|\Psi_{\mathbf{k}\uparrow}\rangle = \tfrac{1}{2}|\phi^d_{\mathbf{k}}(\mathbf{R}_i)|^2 + \tfrac{1}{2}\sum_p \{\alpha_p\, a_{1s}(\mathbf{R}_i)a_{ps}(\mathbf{R}_i)\}\sum_m (c^m_{\mathbf{k}})^2$$

and

$$\langle\Psi_{\mathbf{k}\downarrow}|S_+(\mathbf{R}_i)|\Psi_{\mathbf{k}'\uparrow}\rangle = |\phi^d_{\mathbf{k}}(\mathbf{R}_i)||\phi^d_{\mathbf{k}'}(\mathbf{R}_i)|^* + \Big\{\sum_p \alpha_p\, a_{1s}(\mathbf{R}_i)a_{ps}(\mathbf{R}_i)\Big\}\Big\{\sum_{m,m'} c^m_{\mathbf{k}} c^{m'}_{\mathbf{k}'}\Big\}.$$

α_p is proportional to the exchange integral $\langle 1s, d|g|ps, d\rangle$, using the localized orbitals, divided by the energy difference $E_{ps} - E_{1s}$.

The calculation of the relaxation rate becomes straightforward and we again use the sum rule (A2.2).

For cubic symmetry eqn (8.6) is obtained, the coefficient q of (8.6) having the value

$$q = \tfrac{1}{3}p^2 + \tfrac{1}{2}(1-p)^2.$$

BIBLIOGRAPHY

1. ABRAGAM, A. *The principles of nuclear magnetism.* Clarendon Press, Oxford (1961).
2. ANDREW, E. R. *Nuclear magnetic resonance.* Cambridge University Press (1955).
3. SLICHTER, C. P. *Principles of magnetic resonance.* Harper and Row (1963).
4. ABRAGAM, A. and BLEANEY, B. *Paramagnetic resonance.* Clarendon Press, Oxford (1970).
5. FREEMAN, A. J. and FRANKEL, R. B. *Hyperfine interactions.* Academic Press (1967).
6. KITTEL, C. *Introduction to solid state physics.* Wiley (1963).
7. —— *Quantum theory of solids.* Wiley (1963).
8. LANDAU, L. D. and LIFSHITZ, E. M. *Quantum mechanics.* Pergamon Press (1958).
9. NOZIÈRES, P. *Theory of interacting Fermi systems.* Benjamin (1964).
10. PINES, D. and NOZIÈRES, P. *The theory of quantum liquids.* Benjamin (1966).
11. RODRIGUEZ, S. *Phys. Lett.* **4**, 306 (1963).
12. KUBO, R. *J. phys. Soc. Japan* **17**, 975 (1962).
13. CHARVOLIN, J., FROIDEVAUX, C., TAUPIN, C., and WINTER, J. M. *Solid St. Communs* **4**, 357 (1966).
14. PINES, D. *Solid St. Phys.* **1**, 367 (1956).
15. SAMPSON, J. P. and SEITZ, F. *Phys. Rev.* **58**, 633 (1940).
16. WOLFF, P. A. ibid. **120**, 814 (1960).
17. SCHREIBER, D. S. *Nuclear magnetic resonance and relaxation in solids* (ed. L. V. GERVEN), p. 190. North-Holland (1965).
18. REDFIELD, A. C. and HECHT, R. *Phys. Rev.* **132**, 972 (1963).
19. OBATA, Y. *J. phys. Soc. Japan* **18**, 1020 (1963).
20. OVERHAUSER, A. W. *Phys. Rev.* **92**, 411 (1953).
21. ABRAGAM, A. ibid. **98**, 1729 (1955).
22. LANDAU, L. D. and LIFSHITZ, E. M. *Statistical physics.* Pergamon Press (1959).
23. WOLFF, P. A. *Phys. Rev.* **129**, 84 (1963).
24. KITTEL, C. ibid. **95**, 589 (1954).
25. BLOEMBERGEN, N. and ROWLAND, T. J. ibid. **97**, 1679 (1955).
26. ANDERSON, P. W. and WEISS, P. R. *J. phys. Soc. Japan* **9**, 316 (1954).
27. VAN VLECK, J. H. *Phys. Rev.* **74**, 1168 (1948).
28. RUDERMAN, M. A. and KITTEL, C. ibid. **96**, 99 (1954).
29. BLOEMBERGEN, N. and ROWLAND, T. J. *Acta metall.* **1**, 731 (1953).
30. GOLDMAN, M. *Spin temperature and nuclear magnetism.* Clarendon Press, Oxford (1970).
31. REDFIELD, A. G. *Phys. Rev.* **98**, 1787 (1955).

32. SOLOMON, I. and EZRATTY, J. *Phys. Rev.* **127**, 78 (1962).

33. WINTER, J. M. *Nuclear magnetic resonance and relaxation in solids* (ed. L. VAN GERVEN), p. 61. North-Holland (1965).

34. PROVOTOROV, B. N. *Soviet Phys. JETP* **14**, 1126 (1962).

35. GOLDMAN, M. *J. Physique*, **25**, 843 (1964).

36. MITCHELL, A. H. *J. Chem. Phys.* **26**, 1714 (1957).

37. KAMBE, K. and OLLOM, J. F. *J. phys. Soc. Japan* **11**, 50 (1956).

38. BOLEF, D. I. *Magnetic resonance and relaxation* (ed. R. BLINC), p. 333. North-Holland (1966).

39. GREGORY, E. H. and BÖMMEL, H. E. *Phys. Rev. Lett.* **15**, 404 (1965).

40. BUTTET, J., GREGORY, E. H., and BAIL, P. K. ibid. **23**, 1030 (1969).

41. SCHUMACHER, R. T. and SLICHTER, C. P. *Phys. Rev.* **101**, 58 (1957).

42. —— and VEHSE, W. E. *Physics Chem. Solids* **24**, 297 (1963).

43. HECHT, R. *Phys. Rev.* **132**, 966 (1963).

44. SHIMIZU, M. *J. phys. Soc. Japan* **15**, 2220 (1960).

45. RYTER, C. *Phys. Lett.* **4**, 69 (1963).

46. KJELDAAS, T. and KOHN, W. *Phys. Rev.* **105**, 806 (1957).

47. BROOKS, H. Unpublished. Quoted by G. B. BENEDEK and T. KUSHIDA, Ref. 48.

48. BENEDEK, G. B. and KUSHIDA, T. *Physics Chem. Solids* **5**, 241 (1958).

49. BLOEMBERGEN, N. *Can. J. Phys.* **34**, 1299 (1956).

50. HECHT, R. and REDFIELD, A. G. *Phys. Rev.* **132**, 972 (1963).

51. NARATH, A. and WEAVER, H. T. ibid. **175**, 373 (1968).

52. CALLAWAY, J. and KOHN, W. ibid. **127**, 1913 (1962).

53. RYTER, C. *Phys. Rev. Lett.* **5**, 10 (1960) and private communication.

54. ANDERSON, A. and REDFIELD, A. G. *Phys. Rev.* **116**, 583 (1959).

55. JEROME, D. and GALLERON, C. *Physics Chem. Solids* **24**, 516 (1963).

56. POITRENAUD, J. and WINTER, J. M. *Phys. Lett.* **17**, 199 (1965).

57. —— *Physics Chem. Solids* **28**, 161 (1967).

58. —— Thesis, Orsay (1966).

59. LIEN, W. H. and PHILLIPS, N. E. *Phys. Rev.* **133**, A1370 (1964).

60. GOUSSELAND, O. *J. Physique* **23**, 928 (1962).

61. MAHANTI, S. D. and DAS, T. P. *Phys. Rev.* **170**, 426 (1968).

62. TSERLIKKIS, L., MAHANTI, S. D., and DAS, T. P. ibid. **178**, 630 (1969).

63. POITRENAUD, J. and WINTER, J. M. *Physics Chem. Solids* **25**, 123 (1963).

64. SOGO, P. B. and JEFFRIES, C. D. Quoted in Ref. 27.

65. SCHULTZ, S., SHANABARGER, M. R., and PLATZMAN, P. M. *Phys. Rev. Lett.* **19**, 749 (1967).

66. —— and MONOD, P. *Phys. Rev.* **173**, 645 (1968).

67. NARATH, A. ibid. **163**, 232 (1967).

68. ANDERSON, W. T., RUHLING, M., and HEWITT, R. R. ibid. **161**, 293 (1967).

69. JENA, P., MAHANTI, S. D., and DAS, T. P. *Phys. Rev. Lett.* **20**, 544 (1968).

70. ALLOUL, H. and FROIDEVAUX, C. Private communication.

71. POMERANTZ, M. and DAS, T. P. *Phys. Rev.* **119**, 70 (1960).

72. ALLOUL, H. Thesis, Orsay (1968).

73. ROWLAND, T. J. *Nuclear magnetism in metals*, p. 32. Pergamon Press (1961).

74. —— *Phys. Rev.* **103**, 1670 (1956).

75. SEYMOURS, F. W. and STYLES, G. A. *Phys. Lett.* **10**, 269 (1964).

76. BORSA, F. and BARNES, R. J. *Physics Chem. Solids* **27**, 567 (1966).

77. SHARMA, M. and WILLIAMS, D. L. *Phys. Lett.* **25A**, 738 (1967).

78. KASOWSKI, R. V. and FALICOV, L. M. *Phys. Rev. Lett.* **22**, 1001 (1969).

79. TUNSTALL, D. P. and BROWN, D. *Phys. Lett.* **27**, 723 (1968).

80. KNIGHT, W. D., HEWITT, R. R., and POMERANTZ, M. *Phys. Rev.* **104**, 271 (1956).

81. SHARMA, S. N. and LLEWELYN WILLIAMS, D. *Magnetic resonance and relaxation* (ed. R. BLINC). North-Holland (1967).

82. BLANDIN, A., DANIEL, E., and FRIEDEL, J. *Phil. Mag.* **4**, 180 (1959).

83. FRIEDEL, J. *Nuovo Cim.* Supp. **7**, 287 (1958).

84. WATSON, R. E. *Hyperfine interactions* (ed. A. J. FREEMAN and R. B. FRANKEL). Academic Press (1967).

85. DRAIN, L. E. *Phil. Mag.* **4**, 484 (1959).

86. ROWLAND, T. J. *Phys. Rev.* **125**, 459 (1962).

87. FROIDEVAUX, C. and WEGER, M. *Phys. Rev. Lett.* **12**, 123 (1964).

88. ALLOUL, H. and FROIDEVAUX, C. *Magnetic resonance and relaxation* (ed. R. BLINC). North-Holland (1967).

89. ROWLAND, T. J. *Phys. Rev.* **119**, 900 (1960).

90. —— *Acta metall.* **3**, 74 (1955).

91. MINIER, M. *Phys. Rev.* **182**, 437 (1969).

92. FERNELIUS, N. and SLICHTER, C. P. *Magnetic resonance and radiofrequency spectroscopy*, p. 347. North-Holland (1969).

93. REDFIELD, A. G. *Phys. Rev.* **130**, 589 (1963).

94. AVERBUCH, P., de BERGEVIN, F., and MULLER-WARMUTH, C. *C.r. hebd. Séanc. Acad. Sci., Paris* **249**, 1190 (1959).

95. KOHN, W. and VOSKO, S. H. *Phys. Rev.* **119**, 912 (1960).

96. BLANDIN, A. and FRIEDEL, J. *J. Phys. Radium, Paris* **21**, 689 (1960).

97. MINIER, M. *Phys. Lett.* **26A**, 548 (1968).

98. —— Thesis, Grenoble (1968).

99. ZIMAN, J. M. *Adv. Phys.* **16**, 421 (1967).

100. LACKMANN-CYROT, F. *Phys. Kond. Mat.* **3**, 75 (1964).

101. ROSSINI, F. A. and KNIGHT, W. D. *Phys. Rev.* **178**, 641 (1969).

102. SHOLL, C. A. *Proc. Phys. Soc.* **91**, 130 (1967).

103. BORSA, F. and RIGAMONTI, A. *Nuovo Cim.* **48**, 130 (1967).

104. BONERA, G., BORSA, F., and RIGAMONTI, A. *Magnetic resonance and radiofrequency spectroscopy* (ed. P. AVERBUCH). North-Holland (1969).

105. ODLE, R. L. and FLYNN, C. P. *Phil. Mag.* **13**, 699 (1966).

106. NARATH, A. *Hyperfine interactions* (ed. A. J. FREEMAN and R. B. FRANKEL). Academic Press (1967).

107. JACCARINO, V. *Theory of magnetism in transition metals* (ed. W. MARSHALL). Academic Press (1967).

108. WILSON, A. H. *The theory of metals*, 2nd edn. Cambridge University Press (1958).

109. YAFET, Y. and JACCARINO, V. *Phys. Rev.* **133A**, 1630 (1964).

110. BUTTERWORTH, J. *Phys. Rev. Lett.* **5**, 370 (1960).

111. NARATH, A. and FROMHOLD, A. T. *Phys. Rev.* **139A**, 794 (1965).

112. —— and ALDERMAN, D. W. ibid. **143**, 329 (1966).

113. —— FROMHOLD, A. T., and JONES, E. D. ibid. **144**, 428 (1966).

114. SEITCHIK, J. A., GOSSARD, A. C., and JACCARINO, V. ibid. **136A**, 1119 (1964).

115. BEAL-MONOD, M. T., SHANG-KENG, MA, and FREDKIN, D. R. *Phys. Rev. Lett.* **20**, 929 (1968).

116. CLOGSTON, A. M. and JACCARINO, V. *Phys. Rev.* **121**, 1357 (1961).

117. WATSON, R. E., GOSSARD, A. C., and YAFET, Y. ibid. **140A**, 375 (1965).

118. HEEGER, A. J., KLEIN, A. P., and TU, P. *Phys. Rev. Lett.* **17**, 803 (1966).

119. KASUYA, T. *Magnetism* (ed. G. T. RADO and H. SUHL), vol. 2B. Academic Press (1966).

120. YOSIDA, K. *Phys. Rev.* **106**, 893 (1957).

121. CAROLI, B. and BLANDIN, A. *Magnetic resonance and relaxation* (ed. R. BLINC). North-Holland (1967).

122. BLANDIN, A. and FRIEDEL, J. *J. Phys. Radium, Paris* **20**, 160 (1959).

123. LUMPKIN, O. *Phys. Rev.* **164**, 324 (1967).

124. BRETELL, J. M. and HEEGER, A. J. *Phys. Rev.* **153**, 319 (1967).

125. LAUNOIS, H. Thesis, Orsay (1969).

126. KONDO, J. *Prog. theoret. Phys. Osaka* **34**, 372 (1965).

127. TAKANO, F. and OGAWA, T. ibid. **35**, 343 (1966).

128. JENSEN, M. A., HEEGER, A. J., WELSH, L. B., and GLADSTONE, G. *Phys. Rev. Lett.* **18**, 997 (1967).

129. NARATH, A., GOSSARD, A. C., and WERNICK, J. H. ibid. **20**, 795 (1968).

130. RIVIER, N. and ZÜCKERMANN, M. ibid. **21**, 904 (1968).

131. CAROLI, B., LEDERER, P., and SAINT-JAMES, D. ibid. **23**, 700 (1969).

132. BUTTERWORTH, J. *Proc. phys. Soc.* **83**, 71 (1964).

133. JACCARINO, V., MATHIAS, B. T., PETER, M., SUHL, H., and WERNICK, J. H. *Phys. Rev. Lett.* **5**, 251 (1960).

134. JONES, E. D. quoted by A. NARATH, Ref. 106.

135. PORTIS, A. M. and LUNDQUIST, R. H. *Magnetism.* (ed. G. T. RADO and H. SUHL), vol. 2A. Academic Press (1965).

136. MORIYA, T. *J. phys. Soc. Japan* **19**, 681 (1964).

137. FRANKEL, R. B., HUNTZICKER, J., and MATTHIAS, E. *Phys. Lett.* **15**, 163 (1965).

138. DANIEL, E. *Hyperfine structure and nuclear radiation* (ed. E. MATTHIAS and D. A. SHIRLEY). North-Holland (1968).

139. BENEDEK, G. and ARMSTRONG, A. *J. appl. Phys.* **32**, 1065 (1961).

140. JACCARINO, V., WALKER, L. R., and WERTHEIM, G. H. *Phys. Rev. Lett.* **13**, 752 (1964).

141. KAPLAN, N., JACCARINO, V., and WERNICK, J. M. ibid. **16**, 1142 (1966).

142. DE GENNES, P. G. *Superconductivity of metals and alloys.* Benjamin (1964).

143. COOPER, L. N. *Phys. Rev.* **104**, 1189 (1956).

144. BARDEEN, J., COOPER, L. N., and SCHRIEFFER, J. R. ibid. **106**, 1962 (1957).

145. YOSIDA, K. ibid. **110**, 769 (1958).

146. ABRIKOSOV, A. A. and GORKOV, L. P. *Zh. éksp. teor. Fiz.* **42**, 1088 (1962) (translation *Soviet Phys. JETP* **15**, 752 (1962)).

147. WRIGHT, F., HINES, W. A., and KNIGHT, W. D. *Phys. Rev. Lett.* **18**, 115 (1967).

148. HINES, W. A. and KNIGHT, W. D. ibid. **18**, 341 (1967).

149. DE GENNES, P. G., *Solid St. Communs* **4**, 95 (1966).

150. HEBEL, L. C. and SLICHTER, C. P. *Phys. Rev.* **113**, 1504 (1959).

151. MASUDA, Y. and REDFIELD, A. G. ibid. **125**, 159 (1962).

152. ANDERSON, P. W. *Physics Chem. Solids* **11**, 26 (1959).

153. BUTTERWORTH, J. and MacLAUGHLIN, D. E. *Phys. Rev. Lett.* **20**, 265 (1968).

154. Orsay group on *Superconductivity in magnetic resonance and relaxation* (ed. R. BLINC). North-Holland (1967).

155. GORKOV, L. P. *Zh. éksp. teor. Fiz.* **36**, 1918 (1959) (translation *Soviet Phys. JETP* **9**, 1364 (1960)).

156. —— *Zh. éksp. teor. Fiz.* **37**, 833 (1959) (translation *Soviet Phys. JETP* **10**, 593 (1960)).

157. DOBROSALJEVIC, L. *C.r. hebd. Séanc. Acad. Sci., Paris* **263**, B502 (1966).

158. PINCUS, P., GOSSARD, A. C., JACCARINO, V., and WERNICK, J. H. *Phys. Lett.* **13**, 21 (1964).

159. REDFIELD, A. G. *Phys. Rev.* **162**, 367 (1967).

160. DELRIEU, J. M. and WINTER, J. M. *Solid St. Communs* **4**, 545 (1966).

161. ABRIKOSOV, A. A. *Soviet Phys. JETP* **5**, 1174 (1957).

162. DELRIEU, J. M. *Solid St. Communs* **8**, 61 (1970).

163. EILENBERGER, G. *Phys. Rev.* **153**, A584 (1967).

164. ROSSIER, D. Thesis, Orsay (1969).

165. CYROT, M. *J. Phys. Paris* **27**, 283 (1966).

166. —— Thesis, Orsay (1968).

167. SILBERNAGEL, B. G., WEGER, M., and WERNICK, J. H. *Phys. Rev. Lett.* **17**, 384 (1966).

168. CAROLI, C. and MATRICON, J. *Phys. Cond. Matter.* **3**, 380 (1965).

169. FITE II, W. and REDFIELD, A. G. *Phys. Rev.* **162**, 358 (1967).

170. CYROT, M., FROIDEVAUX, C., and ROSSIER, D. *Phys. Rev. Lett.* **19**, 647 (1967).

171. FEHER, G. and KIP, A. F. *Phys. Rev.* **98**, 337 (1955).

172. DYSON, F. J. ibid. **98**, 349 (1955).

173. DE GENNES, P. G. Rapport SPM no. 469 (French A.E.C.) (1958). Unpublished.

174. RAMSEY, N. F. *Phys. Rev.* **78**, 695 (1950).

175. VANDERVEN, S. and SCHUMACHER, T. R. *Phys. Rev. Lett.* **12**, 695 (1964).

176. LEWIS, R. B. and CARVER, T. R. ibid. **12**, 693 (1964).

177. LAMPE, M. and PLATZMAN, P. M. *Phys. Rev.* **150**, 340 (1966).

178. ALLIS, W. P. *Handbuch der Physik*, vol. 21, p. 395. Springer-Verlag (1956).

179. PLATZMAN, P. M. and WOLFF, P. A. *Phys. Rev. Lett.* **18**, 28 (1967).

180. YAFET, Y. *Solid St. Phys.* **14**, Academic Press (1963).

181. —— *Phys. Rev.* **106**, 679 (1957).

182. BIENENSTOCK, A. and BROOKS, H. ibid. **136A**, 784 (1964).

183. SCHULTZ, S. and LATHAM, C. *Phys. Rev. Lett.* **15**, 148 (1965).

184. —— DUNIFER, G., and LATHAM, C. *Phys. Lett.* **23**, 192 (1966).

185. ELLIOT, R. J. *Phys. Rev.* **96**, 266 (1954).

186. OVERHAUSER, A. W. *Phys. Rev.* **89**, 689 (1953).

187. PINES, D. and SLICHTER, C. P. ibid. **100**, 1014 (1955).

188. ASIK, J. R., BALL, M. A., and SLICHTER, C. P. *Magnetic resonance and relaxation* (ed. R. BLINC). North-Holland (1967).

189. —— —— —— *Phys. Rev. Lett.* **16**, 740 (1966).

190. MONOD, P. Thesis, Orsay (1968).

191. FERRELL, R. A. and PRANGE, R. E. *Phys. Rev. Lett.* **18**, 283 (1967).

192. MOTT, N. F. and MASSEY, H. S. W. *The theory of atomic collisions*. Clarendon Press, Oxford (1965).

193. McELROY, J. A. and HEEGER, A. J. *Phys. Rev. Lett.* **20**, 1481 (1967).

194. SCHULTZ, S. and DUNIFER, G. ibid. **18**, 283 (1967).

195. DUNIFER, G. Thesis, University of California, La Jolla (1968).

196. SILIN, V. P. *Zh. éksp. teor. Fiz.* **33**, 1227 (1957) (translation *Soviet Phys. JETP* **6**, 945 (1958)).

197. SILIN, V. P. *Zh. éksp. teor. Fiz.* **35**, 1243 (1958) (translation *Soviet Phys. JETP* **8**, 870 (1959)).

198. NOZIERES, P. *Polarization, matière et rayonnement*, Presses Universitaires, Paris (1969).

199. PETER, M., DUPRAZ, J., and COTTET, H. *Helv. phys. Acta* **40**, 301 (1967).

200. WINTER, J. M. *Polarisation, matière et rayonnement*, Presses Universitaires, Paris (1969).

201. HASEGAWA, H. *Prog. theor. Phys., Osaka* **21**, 483 (1959).

202. KUBO, R. *J. phys. Soc. Japan* **12**, 570 (1957).

203. NARATH, A. *Phys. Rev.* **162**, 320 (1967).

204. FRADIN, F. Y. *Phys. Lett.* **28A**, 441 (1968).

AUTHOR INDEX

SUBJECT INDEX

actinide group, 113
alkali metal, 16, 77, 154
alloys, 92
aluminium, 90, 108, 140, 145, 154, 165
  alloys, 103–4, 107
aluminium–antimony alloy, 110
aluminium–iron alloy, 125
aluminium–magnesium alloy, 103–4
aluminium-manganese alloy, 125, 127
aluminium–rare-earth alloy, XAl_2, 129
angular momentum, 1
anisotropic Knight shift, 37, 57–8, 89, 91
anisotropy of the relaxation rate, 190
anomalous skin depth, 155
anti-shielding factor, 13, 109
asymmetry parameter, 12

BCS model, 135, 140
beryllium, 89, 154, 164
bismuth, 109
Boltzmann factor, 4, 48
Brillouin:
  function, 134
  zone, 21
broadening of the resonance line, 3, 5–6, 57
  in alloy, 97
  with transition elements, 123
B_0, Landau parameter, 26, 49, 67, 78, 84, 120, 162, 178, 185
B_1, Landau parameter, 162, 170

cadmium, 89, 99, 109
caesium, 77–9, 82, 84–7, 108, 162
chemical shift, 39
coherence length, ξ_0, 138, 146
configuration interaction method, 191
contact hyperfine coupling for contact interaction, 9
copper, 89, 154, 162, 165
  alloys, 100–2, 107
copper–arsenic alloy, 171
copper–chromium alloy, 125, 178
copper–gallium alloy, 171
copper–germanium alloy, 171
copper–iron alloy, 126, 178
copper–manganese alloy, 121, 129, 171, 178
copper–nickel alloy, 125
copper–zinc alloy, 171
core-polarization:
  effect, 11

Knight shift, 115
  relaxation rate, 117–8, 190;
correlation:
  function, 49, 62, 67, 186
  coefficient, ϵ_{ij}, 63, 65, 67, 82–4, 141
creation operator, 18, 142
critical field, H_c, 139
  in type II superconductors, H_{c_1}, H_{c_2}, 139, 146
cross-section for spin-flip scattering, 168
cross-relaxation rate, 183
cyclotron resonance, 154
  frequency, 24, 160

d-band, 113–14, 128; see also d-electron
d-electron, 10, 14, 124, 131, 189
d-phase shift, 124–5, 133, 171
d-shell, 10, 113; see also d-electron
De Haas–Van Alphen effect, 24
Debye temperature, 136
defects, 92
demagnetizing field, 38
density of states, 17, 23–4, 89, 90, 114–15, 120
  in a superconductor, 142
destruction operator, 18, 142
diamagnetic susceptibility, 24, 115
diffusion:
  effect, 58, 80–1, 87
  of the electron, 154, 158
  length, 156
  in liquid metal, 109–10
dipolar:
  hyperfine coupling, 9, 10
  interaction, 1–6, 7, 87–8
  Knight shift, 36
  relaxation rate, 43, 82, 189
divalent metal, 89, 154

effective mass, 26, 84–5, 87, 162
electric:
  field gradient in alloy, 105–6
  hyperfine interaction, 11
  quadrupole moment, 12
electron–electron interaction, 20, 48, 50, 82
electronic:
  density at the nucleus, 77–9, 80, 85
  gyromagnetic ratio, g, 1, 152, 161–2
  line width, 164
  mean free path, 140, 150, 155